INTO THE CLEAR BLUE SKY

THE PATH TO RESTORING
OUR ATMOSPHERE

ROB JACKSON

SCRIBNER

NEW YORK   LONDON   TORONTO   SYDNEY   NEW DELHI

Scribner
An Imprint of Simon & Schuster, LLC
1230 Avenue of the Americas
New York, NY 10020

Certain names and characteristics have been changed.

Copyright © 2024 by Robert B. Jackson

All rights reserved, including the right to reproduce this book
or portions thereof in any form whatsoever. For information,
address Scribner Subsidiary Rights Department,
1230 Avenue of the Americas, New York, NY 10020.

First Scribner hardcover edition July 2024

SCRIBNER and design are trademarks of Simon & Schuster, LLC

Simon & Schuster: Celebrating 100 Years of Publishing in 2024

For information about special discounts for bulk purchases,
please contact Simon & Schuster Special Sales at 1-866-506-1949
or business@simonandschuster.com.

The Simon & Schuster Speakers Bureau can bring authors to
your live event. For more information or to book an event, contact
the Simon & Schuster Speakers Bureau at 1-866-248-3049
or visit our website at www.simonspeakers.com.

Interior design by Laura Levatino

Manufactured in the United States of America

1 3 5 7 9 10 8 6 4 2

Library of Congress Cataloging-in-Publication Data has been applied for.

ISBN 978-1-6680-2326-6
ISBN 978-1-6680-2328-0 (ebook)

For Sally, whose stunning ceramic books inspire real ones, and for Robert, David, and Will.

And for the many lab members, friends, students, and mentors I've learned from, especially Martyn Caldwell and Hal Mooney.

If you want a happy ending, that depends, of course, on where you stop your story.

Orson Welles with Oja Kodar, *The Big Brass Ring*

I don't know, Bobby. You have to believe that there is good in the world. I'm goin to say that you have to believe that the work of your hands will bring it into your life. You may be wrong, but if you don't believe that then you will not have a life.

Cormac McCarthy, *The Passenger*

We set this house on fire forgetting that we live within.

Jim Harrison, *To a Meadowlark*

CLIMATE STASIS

We waited, delayed, hesitated,
procrastinated, prognosticated, obfuscated on thin ice,
dawdled, stalled, free-for-alled, protocolled collect,
dropped balls, stonewalled, mitigated communities,
modeled, dallied, rallied, death valleyed,
risked, bet, staked, gamed, gambled, generally wagered all
and broke bad, faith, dreams, records, ranks, and the bank.

Rob Jackson, *Light*

CONTENTS

PROLOGUE

"Will I make it?"

"Will I make it to the end of my lifetime?" The question came out of the blue—the blue cheese of a dessert plate. We'd just reshuffled our dinner chairs like playing cards and introduced ourselves to new table-mates.

Ashley Martin asked me the question after hearing what I did for a living. An assistant professor in Stanford's Graduate School of Business at the time, Ashley ostensibly had everything: youth, intelligence, income, and prestige. Why was *she* of all people pondering this question? But then again, why wouldn't someone of her generation ponder it?

I'd grown used to being asked other questions: "Should I have my first child today?" ("Yes"—to keep things simple.) "What can I do to fight climate change?" That question has no quick answer and, in part, motivates this book.

After Ashley asked me, "Will I make it to the end of my lifetime?" she added a second gut-puncher: "And will I be okay?"

I can't say why, but her vulnerability left me fighting back tears. My first thought was "This is not your fault, and you shouldn't have to bear this grief."

But Ashley is one of the lucky ones. She lives in one of the wealthiest cities in the wealthiest country on Earth. She certainly could lose her home, her loved ones, and yes, even her life, to climate change—many people in relatively rich nations already have—but that isn't likely. Ashley and I and many of you reading this book have more resources than most to adapt to the climate change that is only getting worse. Other people and species on Earth aren't so fortunate.

Faria Khan's question was even more pointed. She emailed me soon after my dinner with Ashley.

Dear Professor Rob,

Hope you are doing well and with the semester starting, you are working on exciting new research. I wanted to learn something from you. I hope you are following up on the floods that Pakistan is experiencing this year. It has wreaked havoc in the country.

I wanted to learn from an expert like you—Pakistan only contributes towards 1% of global carbon emissions yet it is one of the countries most affected by climate change. What can we do to reduce such instances in the future in the country? Normally, 4 glaciers melt per annum while this year 16 glaciers from the Himalayan and Karakoram ranges melted and wreaked havoc in the whole country. I want to help the policy makers in Pakistan better prepare for such changes in the future. Can you please identify the measures which we can take to help control such drastic effects?

Thanks in advance, looking forward to hearing from you.

Kind regards,
Faria

How do you respond to a person whose country is one-third submerged by monsoons turbocharged by climate change? Sherry Rehman,

Pakistan's climate minister, said the flooding "exceeded every boundary, every norm we've seen in the past."

How can you respond to the question "What can we do to reduce such instances in the future?" And who is the "we"? Pakistan contributes less than 1 percent of global fossil carbon dioxide pollution, and its per capita carbon emissions are a tenth or less of those in the United States. Why must Pakistanis suffer the effects of climate change that they did so little to cause? This imbalance is what Rehman calls "climate colonialism" that penalizes "low polluters who are paying the costs for decades of fossil-fuel development by rich countries."

Any hopeful book about climate solutions, such as this one, should discuss exciting new technologies that can help us provide zero-carbon energy to everyone from Pakistan to Europe and the United States. It must also grapple with issues of equity, justice, and greed.

Ashley's grief is unfair, Faria's even more so. For Ashley, Faria, and the millions with questions I struggle to answer, this book—and the hope within—is for you.

INTRODUCTION

Michelangelo stands above me.

Shoulder-bumping my way through the white marble entrance to the Vatican Museum, I flee the human tide and enter the inner sanctum of museum headquarters. I knock on a thick wooden door at the office of Dr. Vittoria Cimino, director of the Vatican Museum's Conservator's Office. Unlike Dr. Cimino I am neither perspiration-free nor pulled together, having just walked from the metro during a heat wave that marks Rome's hottest temperature ever in a record-breaking summer where sixty thousand Europeans will die of heat.

I'm visiting Cimino and the museum to learn about conservation and restoration—the balance of inspiration, perspiration, technology, and faith needed to sustain treasures for centuries. Hundreds of red and green binders line her office wall—long-term records of care for the more than one hundred thousand priceless works she and her staff curate. She is also an expert on the Sistine Chapel, which recently underwent a decades-long effort to restore the lost colors of Michelangelo's *The Last Judgment* and other frescoes.

At the start of our conversation, she wants to make sure I note the distinction between conservation and restoration. "We use preventive conservation to maintain everything in a good condition—the environment, the buildings, and all works, from the masterpiece to a little piece of art," she says.

"Restoration is different. Restoration focuses on one object at a time and requires a huge investment: economic, administrative, and human.

We conserve and restore today because maybe people will have better technology in the future."

Vittoria Cimino's perspective applies to more than restoring priceless works of art. Restoring the atmosphere—my dream as a climate scientist and the reason I'm writing this book—combines the dedication Vittoria applies to her work with long-term environmental thinking: decadal, multigenerational, even "cathedral" thinking that sustains stewardship over centuries for the benefit of descendants.

Restoring the atmosphere to preindustrial levels of greenhouse gases and other pollutants must invoke the same spirit and philosophy used to restore endangered species and habitats to health. The U.S. Endangered Species Act doesn't stop at saving plants and animals from extinction. It mandates recovery. When we see gray whales breaching on their way to Alaska each spring, grizzly bears ambling across a Yellowstone meadow, bald eagles and peregrine falcons soaring on updrafts, we celebrate life and a planet restored. Our goal for the atmosphere must be the same.

Our timing is fortuitous. The United Nations named the 2020s the "Decade on Restoration" to repair degraded ecosystems, combat climate change, and safeguard biodiversity, food, and water supplies. What better way to realize success than by returning greenhouse gases such as methane and—much later—carbon dioxide to preindustrial levels? That's what I mean by "restoring the atmosphere" for a given greenhouse gas. We must cut emissions to stay below global temperature increases of 1.5 or 2°C. But such abstract targets aren't motivating people to change habits and behaviors. We've sprinted to the precipice of 1.5°C—putting lives and livelihoods in peril—and need a more powerful narrative for success.

I ask Cimino how far in advance her office plans their care—decades or centuries into the future? She laughs. "We are realists. We act for the following decade, and we hope the following decade will act for the decade after that."

I recall her words later as I enter the chapel. *The Last Judgment* fills the altar wall in whirling blue-sky motion. People rise on the left and fall on the right, their fates captured in their eyes, upturned or down-trodden. Christ turns the wheel—one arm up, one arm down—from the fresco's upper center.

The intensity of the blue sky uncovered by the decades-long resto-ration is breathtaking. Michelangelo's ultramarine-blue pigment made of lapis lazuli—now restored and visible for the first time in centuries—originally cost the Vatican a queen's ransom. The same mineral adorns King Tut's blue-and-gold death mask.

Scanning *The Last Judgment* systematically from top to bottom, I find four or five small dark rectangles that look like a child snuck in, scampered up scaffolding, and colored patches of fresco with black crayon. I gasp at their contrast. They are the conservator's *testimoni* that Vittoria described, "evidence of what used to be, so people will re-member." One narrow rectangle in the lower left floats as dark smog suspended in an azure sky.

To reveal the original colors, restorers needed to remove dirty oils and resins without damaging the plaster or pigments underneath. They placed gauze-like sheets moistened with solvents on the frescoes one patch at a time. When they peeled off the fabric, the grime vanished. "It was a kind of miracle," Cimino said about seeing Michelangelo's colors and faces restored in waves along the walls of the chapel like sunshine emerging from a storm cloud.

Centuries of grime wasn't the only thing creating damage that re-quired restoring, however. More than a decade ago, Vittoria Cimino's office discovered that parts of the frescoes were turning powdery white. Scientists examined the deposits and found them to be calcium carbon-ate. Extra carbon dioxide and humidity from people's breath had com-bined to form a salt—calcium bicarbonate—that left its traces on the frescoes wherever condensed water evaporated. The chapel's original air handling system was designed for only seven hundred visitors at a time, but thousands of people were now crowding the chapel daily. Carbon dioxide concentrations inside spiked to five times the levels outdoors.

The Vatican faced a choice. To protect the frescoes, they could restrict the number of visitors or build a more sophisticated air handling and purification system. They chose a new system that could control everything from temperature and relative humidity to levels of pollutants that included ozone, sulfur dioxide, dust, and carbon dioxide.

It's easier to reduce carbon dioxide concentrations in a chapel than across the globe. Our fossil fuel use is still pumping forty billion tons of carbon dioxide pollution a year into the air. Unfortunately, even if we eliminated all of those emissions today, restoring the atmosphere for carbon dioxide won't happen anytime soon. There's too much of it in the air—an extra trillion tons that, left to its own devices, will warm the Earth for millennia. Carbon dioxide is responsible for about half of recent warming. That makes CO_2 Public Enemy #1 for climate and the key target for emissions reductions.

But methane is hot on carbon dioxide's heels as Public Enemy #2. Methane has caused another third of recent warming on top of the damage from CO_2. Much of my research today centers on methane because of its greater potency and shorter lifetime in the atmosphere, factors that make it a prime target for reducing global warming sooner rather than later.

If we want to reduce greenhouse gas warming over the next decade or two, reducing atmospheric methane concentrations through emissions reductions and atmospheric removal is the best—and perhaps only—lever we have to shave peak temperatures and reduce dangerous weather disasters and heat waves like the one I experienced during my visit to Rome. In fact, methane is the only major greenhouse gas for which we could restore the atmosphere "in a lifetime," a key part of my dream.

Methane is deadly. Pound for pound, it's eighty to ninety times more potent than CO_2 at warming the Earth in the first two decades after release, and it's roughly thirty times more potent over a century. Its concentration has doubled over the past century, too, a much more rapid increase than for CO_2. And methane concentrations are accelerating faster today than at any time since record-keeping began.

Methane is mysterious. We're still trying to understand why methane concentrations are surging. It could be that warming temperatures are boosting emissions from tropical wetlands, the largest natural source. It could be that more methane is belching from cows, oil and gas fields, coal mines, landfills, and thawing Arctic permafrost. Scientists measure each of these sources—I work extensively in oil and gas fields and in wetlands, for instance. To understand methane's rise, though, we need to know how quickly methane enters the atmosphere *and* leaves it. Methane concentrations increase if emissions rise, of course, but also if methane lasts longer in the air through chemical reactions with other pollutants. Both things appear to be happening today—higher emissions and slower destruction of methane in the atmosphere.

Methane is neglected. The first-ever major methane agreement—the Global Methane Pledge—was initiated only a few years ago. More than a hundred and fifty nations have joined the United States and European Union in committing to cut methane emissions at least 30 percent by 2030, through reducing energy-based and agricultural emissions.

Methane is also malleable—a plus for climate action. It's cleansed from the air naturally only a decade or so after release. Because of this shorter lifetime, if we could eliminate all methane emissions from human activities, like agriculture, waste, and fossil fuels—a *big* if—methane's concentration would return to preindustrial levels within only a decade or two.

Restoring methane to preindustrial levels would save 0.5°C of warming and could happen in your lifetime—and mine. That's my holy grail as an environmental scientist and as chair of the Global Carbon Project (GCP), one of the premier scientific organizations tracking the world's greenhouse gas emissions. Comprising hundreds of scientists, we in the GCP measure methane emissions from both natural and human-made sources as well as how fast "sinks" remove methane from the atmosphere—the natural processes that clean greenhouse gases from our air. We do the same for the greenhouse gases carbon dioxide, nitrous oxide, and hydrogen. Knowing where greenhouse gases come from gives us the power to reduce fossil pollution.

That job is urgent because fossil fuels are making the Earth unlivable and—even ignoring climate change—they are deadly. In the United States, particulate pollution from coal and cars kills more than a hundred thousand Americans a year, mostly through heart and lung diseases, far more than are murdered, die in traffic accidents, and drown combined. Globally, one in five deaths is attributable to burning fossil fuels—ten million senseless deaths a year—when cleaner, safer fuels are already available.

Remarkable examples show how reducing pollution over decades has improved our health and economic well-being. Lead levels in the blood of young children in the United States have dropped 96 percent since the phaseout of leaded gasoline; the value of this phaseout globally is estimated to be $2.5 trillion per year. Protection of the ozone shield through the safeguards of the Montreal Protocol eliminated billions of skin cancers and millions of cataracts and deaths worldwide. Half a century of action through the U.S. Clean Air Act currently saves hundreds of thousands of lives a year at a thirtyfold return on investment. Workers are healthier and more productive. We breathe easier and pay lower medical expenses related to air pollution.

There's a major difference between most air pollutants and greenhouse gases, though: air pollutants typically disappear quickly—often within weeks or days—after we stop releasing them. Greenhouse gases don't. Our air brims with legacy emissions of greenhouse gases, and emissions reductions alone won't always be enough to restore the atmosphere fully. That climate change is only a "century-timescale phenomenon" is a dangerous misconception. Absent intervention, elevated carbon dioxide concentrations in our air will remain for tens of thousands of years—longer than people have practiced agriculture.

This is one area where my thinking has evolved. I used to believe that talk of hacking the atmosphere—sucking greenhouse gases back out of the air—distracted us from the real job of cutting emissions. It does, but inaction in the lost decade of the 2010s convinced me that to maintain a safe and livable climate, we must develop technologies to remove carbon dioxide, methane, and other gases directly from the air using everything from the microbes, trees, and factory arrays described in this book. And we'll need to do it at industrial scales, running the coal industry in reverse.

There is no one path to restoring the atmosphere. There are many paths—just as there are many roads for driving between Washington, D.C., and San Francisco. To have any chance at a safe climate, though, we know the world must address both carbon dioxide and methane. We need to zero out carbon dioxide emissions, the focus of most climate action today. The unappreciated good news is that every time we cut carbon dioxide emissions from burning oil, gas, and coal, we also cut methane emissions from drilling, mining, and transporting them. Similarly, we need to eliminate direct methane sources from fossil fuel use and agriculture, combining emissions cuts with removal technologies to restore the atmosphere.

These goals structure my book. The cheapest, safest, and only sure path to a safe climate starts with slashing emissions. I begin with some of the largest emissions sources in our lives that we control directly—at home and from our food and vehicles. Transportation and housing contribute at least half of total carbon emissions for people across the United States and United Kingdom, with housing being the largest single emissions source for people in the UK. I turn next to select industries such as steel production because emissions from steel, cement, and a few other industries individually are larger than the emissions from most of the world's countries. We can't decarbonize such industries on our own through personal actions; only systemic change will suffice. The need for systemic change also underpins the pipelines running below our streets and sidewalks that supply gas to our homes—with one critical difference. Few of us could eliminate steel or cement use in our lives, but most of us could choose to end our natural gas use by switching to electric appliances. I finish the emissions section by highlighting gases that are being restored to preindustrial health. In the case of ozone-depleting halocarbons, we already are restoring the atmosphere. Doing so has cooled the planet and saved millions of lives.

The second section focuses on removal or "drawdown" technologies. Because wealthier nations have already dumped so much fossil fuel pollution into the air, we will need to do more than just cut emissions. We will need to actively remove some greenhouse gases from the atmosphere. I begin with examples for carbon dioxide removal, industrially and using nature. Such approaches are more expensive than cutting emissions but are already operating today. I turn to methane removal next. Methane removal is far newer—more concept than reality today—but it's a field of research I am pursuing and have helped to create. I believe the world will need methane removal to offset ongoing emissions from food production and from perturbed natural ecosystems such as Arctic permafrost and tropical wetlands that already show signs of releasing more methane as the Earth warms. Once freed, a genie does not go back in the bottle easily.

I finish the book with stories of success—from fossil fuel companies transforming their businesses to individuals changing the world in their homes, neighborhoods, and beyond. Our journey ends in the Amazon. There, even one of the wildest and remotest habitats on Earth is threatened by climate change. We'll need to act quickly to save it and ourselves. The vision of an atmosphere restored should help.

Drifting front to back in the Sistine Chapel, I pass below the nine famous scenes from Genesis that Michelangelo painted on the ceiling, from Adam and Eve to Noah. In the center, God floats in the sky, right arm and finger stretching for Adam. Adam reclines along a line of hills, his left elbow resting languidly on his knee.

INTRODUCTION

Much has been made of the most famous gap in art—two fingers close but not quite touching—a metaphor for the gap between God and man, between knowledge and responsibility, between responsibility and action, between inaction and judgment.

I exit the chapel reluctantly before ducking back in for a final look. I picture scaffolding, a decade of restorers just starting to cleanse centuries of grime from the frescoes one puzzle piece at a time. In the beginning, their task must have seemed as daunting as ours today: wiping centuries of fossil fuels from our lives one car, one cow, one coal plant at a time.

The day of my visit was Vittoria Cimino's last workday before retirement. Reflecting back on her career, she offered me final guidance: "Restoration is patience."

Part I
EMISSIONS FIRST

1

Fair Shares

From "Code Red" reports highlighting climate calamities to gleaming white wind turbines to hydrogen in green or blue hues, a rainbow of climate consequences and technologies unfurls before us. Before exploring new technologies and climate solutions, however, we must grapple with issues of fairness and equity, consumption and demand, ethics and justice. New technologies will help us restore the atmosphere to preindustrial health, but we can't simply build our way out of climate change—no matter how green new technologies may be.

We live in a time when the top 1 percent of the world's population contributes more fossil carbon emissions than half the people on Earth. In class I call our time, somewhat facetiously, the *Myocene*, the era in which the top tenth of humanity emits half or more of global carbon pollution.

What do global hyper-consumers in the United States and elsewhere gain by consuming so much? Not much, my research suggests. I led a recent analysis of health, well-being, and energy use across 140 countries. Energy use was linked positively to every aspect of well-being we examined, including infant mortality, life expectancy, air quality, food supply, happiness, and prosperity—but only to a point. People's lives improved substantially with access to more energy, but the benefits plateaued at energy uses well below the global average. People in energy-rich countries fared no better than people in countries who used many times less energy per person. Some of us could apparently use far less energy without compromising our health, happiness, or prosperity—and possibly improving our mental and physical health by reducing stress and decluttering our lives.

Everyone needs access to a minimum threshold of energy for a decent quality of life, but using four times the global average nets U.S. residents nothing, at least for the metrics of well-being we examined. Headlines about our research read "How much energy powers a good life? Less than you're using, says a new report" (NPR's *All Things Considered*), "World Doesn't Need More Energy to End Poverty" (Bloomberg News), and "Where the Energy Link to Well-Being Starts Fraying" (*Axios*). Equity in energy—and everything—girds climate solutions.

I often receive emails and phone messages after our studies are published, sometimes nice and sometimes threatening. Regarding this study, I received many, including, "I recently learned of a comment from you, that Americans should be content to live on a quarter of the energy they do now. I gave it some thought and then considered it to be an unrealistic, if not bizarre, expectation. For my wife and I to reduce our power consumption by that much would require turning off our electricity the majority of the time. Are you living by this credo? If so, I would be interested to know how it's working out for you without a refrigerator, air-conditioning, central heating system, etc, for 18 hours of the day? I invite your reply."

I don't typically reply, but I did in this case: "Thank you for reaching out. I aspire to this credo, and as I improve at home it's working out pretty well. Most of our energy use comes from cars, planes, and homes. One could choose to not drive one's car 5 days a week or choose instead to live close enough to work to commute by bike, as I do. If that's not possible, one could carpool or take mass transit at least a few days a week. At home, I traded in my gas furnace for a heat pump that is substantially cleaner and more energy efficient. I don't have air conditioning and we keep our winter thermostat at 62, as people all across Europe do. When I had gas appliances, my household gas use was one-quarter of the U.S. average. That's not too bad and it didn't destroy my life. Now I use no gas at all. It's possible and, I would argue, has made me and my family healthier."

My research also shows that today's global energy consumption—if distributed more equitably—could allow everyone on Earth to realize a good quality of life. That includes at least a billion people stuck in energy poverty who lack access to reliable electricity. They need to use more energy, not less.

Climate solutions will require fairer energy use—either at lower levels of consumption than in the United States or, if absolutely necessary, at U.S. levels. But what would the world look like if everyone consumed at that higher rate?

A fifth of the billion and a half gasoline-powered vehicles on Earth are in the United States, with almost one passenger vehicle per person. Europe has one vehicle for every two people, South America one for five, Asia and Africa one for every seven and twenty people, respectively. If eight billion people on Earth owned cars at the U.S. rate, the world would have seven billion vehicles, almost five times the number today. No matter how "green" those new vehicles might be—electric vehicles (EVs) or otherwise—adding five billion more won't make the world more sustainable by any definition.

More electric cars and trucks mean more transmission lines riving the landscape, more lithium mining for batteries, and more spontaneous lithium battery fires. More biomass burning for electricity means more logging, truck trips, and wilderness roads. More solar panels mean increasingly large land areas needed for harvesting electrons, less land for biodiversity and natural areas, and more mining of rare metals such as gallium and indium. What's true for renewables applies to fossils, too—more fossil fuel energy means more drilling, mining, spilling, and refining. Climate solutions start with consuming less.

When Faria Khan asked me what was fair about Pakistan contributing only 1 percent of global carbon dioxide emissions yet being one of the countries most hurt by flooding and climate change, the only possible answer was "nothing." My Australian colleague Pep Canadell, executive director of the Global Carbon Project, and I wrote a recent opinion piece in the *New York Times* titled "What's 'Fair' When it Comes to Carbon Emissions?" Our two countries—the United States and Australia—together emit one-sixth of the world's fossil fuel pollution with less than one-twentieth of its population. Despite falling emissions for more than a decade due to declining coal use, carbon emissions per person in both countries remain some of the world's highest.

Explicit or implied, issues of fairness and environmental justice underpin all discussions of climate solutions in the following pages—among richer and poorer nations and among richer and poorer people within richer nations. Fairness is about how much we use and who bears the harm from pollution. People like Faria Khan certainly bear disproportionate harm, as do people in poorer communities worldwide.

To understand these concerns better, I turned to environmental justice expert Catherine Coleman Flowers. She founded the Center for Rural Enterprise and Environmental Justice in 2002 to advocate for "grassroots-led solutions in policy and technology." A passionate champion of climate action, she is also an international expert on sanitation and waste. When she was awarded a MacArthur "genius grant" in 2020, her citation praised her for "Bringing attention to failing water and waste sanitation infrastructure in rural areas and its role in perpetuating health and socioeconomic disparities." She is a fierce proponent of environmental justice broadly and a tireless advocate for poorer communities in the southern United States—particularly the two million people whose homes lack indoor plumbing, septic systems, and connections to sewer systems.

"Environmental justice" is defined by the U.S. Environmental Protection Agency (EPA) as "the fair treatment and meaningful involvement of all people regardless of race, color, national origin, or income, with respect to the development, implementation, and enforcement of environmental laws, regulations, and policies." It requires the same protections from harm for all people plus equal access to the decision-making process to guarantee everyone's right to live and work in a healthy environment.

Coleman Flowers and I had met a few years before, when we spoke together on a panel during the United Nations Climate Change Conference in Glasgow, Scotland (referred to as COP26, which stands for conference of the parties). She sat before me in a blue business suit, sharing a smile as infectious as the solutions (and justice) she promotes. I sought her thoughts on the development of new clean technologies, including how and where to deploy those technologies (and distribute their pollution) more equitably across communities than happened during the fossil fuel era.

Coleman Flowers describes herself as a "teacher activist." "I'm an educator," she began. "I was a high school teacher and a junior high school teacher in social studies, a subject I think is sorely needed right now. Social studies and the liberal arts provide the understanding for everything that allows us to act collectively as a society. I've been able to draw upon those skills to not only educate people, but to educate myself, because part of teaching is learning, and I've learned a lot from the people I've encountered."

In Lowndes County, Alabama, for instance, where she grew up, rural—mostly Black—residents are denied access to city or county sewer systems and often can't afford a septic system. Many have no choice but to "straight-pipe" sewage from their toilets onto the grass outside their homes. Children play in this grass; mosquitoes breed in the stagnant pools. Residents fall prey to diseases that are normally found only in the poorest tropical countries. Coleman Flowers fights for friends like Pam Rush, who lived for years with sewage pooling outside her door, her friend "Shar," and countless others who want their families to be able to play safely outside.

That's why a 2017 study that Coleman Flowers organized with doctors at the Baylor School of Medicine made waves when it showed that more than a third of residents they tested in Lowndes County had hookworm, a disease believed to have been eradicated from the United States in the 1980s.

Hookworms affect more than half a billion people worldwide, burrowing under their skin and moving to their intestines. Although primarily tropical, hookworms used to be found across the warm, humid southeastern United States, from east Texas through Virginia. Proper sanitation eliminated most cases because a common path for infection is for people to walk barefoot across ground contaminated by sewage. That hookworm and related pathogens have reappeared in the United States is a travesty of poverty and systemic racism.

It's a travesty of climate change, too. Hookworms are restricted in range by cold temperatures. They tolerate, even thrive in, warmer areas. Not surprisingly then, many species are expanding their ranges. Scientists call this process "global worming."

Coleman Flowers realized that the wetter weather and more extreme storms already happening in the southern United States made it harder for sewage to drain into the soil and more likely to spread across yards during heavy rain. "Affordable sanitation is a right," she told me. "People can't afford to fight climate change and poverty at the same time."

Beyond our homes, Coleman Flowers champions greater equity in the planning and siting of new energy infrastructure and climate solutions. "I don't believe in sacrifice zones," she said, "where someone thinks it's okay to sacrifice one group of people for the good of another." A sacrifice zone is an area that bears exceptional environmental hazards from toxic or polluting industries located nearby—some of the most poisoned and polluted places on the planet. The health and safety of the people in these communities, who are typically lower-income residents and people of color, are sacrificed for the economic gains and prosperity of others located elsewhere.

"Sacrifice zones are not the way to equity," Coleman Flowers continued. "And equity is an issue for both water quality and air pollution. A lot of these problems have been created—and made worse—because of structures put in place that exploited people or that were built upon racist principles." For instance, people of color comprise 45 percent of residents living within two miles of commercial hazardous waste facilities in the United States. People living within six miles of a coal or natural gas power plant are also far more likely to be poor and of color.

Sacrifice zones and environmental equity are global concerns. In Europe, France operates more than one hundred municipal waste incinerators that produce ash and slag and emit dioxins, heavy metals, and particulates. Such facilities are disproportionately located near where immigrants live. In the UK, scientists who performed an analysis of air quality and poverty concluded that "the poorest tend to experience the worst air quality." People living in the poorest 10 percent of neighborhoods breathed far more nitrogen dioxide (NO_2) pollution generated by fossil fuel combustion than the national average. (NO_2 is the same asthma trigger we will measure indoors from combustion by gas stoves in Chapter 2.) The clean-energy transition is an opportunity to remedy injustices in pollution exposures and to improve water, air, and soil quality for all.

"We have to be mindful we aren't just powering on, instead of realizing that some structures must be dismantled," Coleman Flowers said. "Otherwise, we'll never have the type of equitable solutions that will impact all of us—that will yield the kind of successes we're seeking—which are to reduce carbon and hopefully hold things to where we don't have any more climate chaos than we've already created."

Coleman Flowers experienced some personal climate chaos a few weeks before we spoke. She had recently moved to Madison, Alabama, just south of the Tennessee border. "It was eighty degrees on New Year's Day," she said, "which set a record high temperature for the state. The next day tornadoes touched down, one of them within a few miles of my house. And the next day we had snow!

"The dramatic swings we're seeing are crazy," she continued. "Out there [in California] you're getting the wildfires and potentially the droughts, but here we've gotten lots of water." The southeastern United States, where Coleman Flowers lives, now has two times more drenched days with three or more inches of rain than it used to. The extra water is a problem on top of water tables that are already high and septic systems already failing. "We're going to see even more septic failures with sea level rise, because the groundwater tables are getting even higher," Coleman Flowers said. "And then we have all this extreme weather because we have warm currents coming off the Gulf, and colder air coming from the north. It collides here, and we get turbulent weather."

"People feel swamped by climate change," Coleman Flowers said. "We all feel overwhelmed sometimes. Yesterday, I got a little overwhelmed, so I had to stop everything and bake some gingerbread."

Liking Catherine Coleman Flowers comes easily. "Part of my optimism comes because I'm a Black woman that grew up in the South, along the Selma-to-Montgomery march trail," she said. "The reason I'm optimistic is because I never allowed my conditions or what people put upon me to keep me in a box. And likewise, I think when it comes to not only environmental justice, but climate justice, too, I think we will get to where we need to go, because ultimately the deciding factor is Mother Nature. And she can change minds."

"Mother Nature" is already changing minds, or at least making her presence felt, because we're already paying for climate inaction. The National Oceanic and Atmospheric Administration (NOAA) has been collecting data on billion-dollar weather disasters in the United States since the early 1980s. Such disasters cost Americans $165 billion in damages and killed 474 people in 2022. Adjusted for inflation, billion-dollar disasters happen twice as often today as they did a few decades ago. Americans are also paying $50 to $100 billion more a year for them.

Coleman Flowers continued. "We can't predict what Mother Nature will do or where she's going to do it, but that's the deciding factor. I think that's also the unknown factor because it's not something we can control. What we *can* control is how much worse things get."

While things get worse (before they get better), people in lower-income neighborhoods and nations will keep paying more than their share. Half of all U.S. households facing energy insecurity are African American, despite being only 15 percent of the U.S. population. People in lower-income areas worldwide are more vulnerable to floods, storms, storm surges, droughts, and other weather and climate extremes.

"Many people who didn't believe in climate change are suffering today," Coleman Flowers said. "They don't know what to do. Some people are pretending it's not real, but others, when it gets close to them and they see the suffering we could have prevented, they want to do something."

In the sacrifice zones and cancer alleys of the world, perhaps a different family of solutions is needed, solutions where we start by *asking* people for their ideas of what they need and want. Coleman Flowers promotes the voices of those who've borne more than their share of pollution: "They have the type of knowledge we need to connect with the people in academic halls. We need real solutions that can work locally. That's what I've learned coming from a rural community—there's a lot of indigenous knowledge that could provide answers we don't tap into, because our assumption is that because people have not left the area, they don't know. They may not have had the chance to travel the world, but they do have knowledge to help us find solutions quicker if we would just have them sitting at the table."

2

Home on the Range

R ed or black? High or low?

You don't always know when dangerously high pollutants will appear when you light a gas stove. They don't, and then they do—often enough for games of roulette to come to mind—except the wheel or chamber is the knob on a gas burner or oven.

Our homes are an apt place to begin examining greenhouse gas pollution because they comprise one of the biggest sources in our lives. Not including the emissions from producing food, our homes and buildings contribute a third or more of all greenhouse gas emissions in the United States and Europe, primarily through electricity use and the need to heat our water, living spaces, and meals.

My lab and I have visited hundreds of homes with environmental justice groups in lower-income and other neighborhoods to measure methane and carbon dioxide emissions and air pollution arising from burning fossil fuels indoors. Our work is a chance to heed Catherine Coleman Flowers's call to bring everyone to the table—we've sat at kitchen tables in hundreds of homes across Australia, the United States, China, and Europe. Gas use ranges from 40 percent of homes in Australia and roughly half of U.S. homes to nearly all homes in the Netherlands, where a transformation to cleaner electric stoves is underway. In each country we test for toxins in the natural gas entering homes and the pollutants released by stoves when lit. While in people's homes, we also listen to their concerns and their ideas for solutions.

Natural gas is the fastest-growing fossil fuel in the world. It's cleaner than coal but dirtier than almost everything else in use today (unless you're one of the billion people still cooking over wood indoors). The world is approaching eight billion tons of carbon dioxide pollution annually from gas use alone. That level of carbon pollution from only one fossil fuel is incompatible with a world where we stabilize global temperatures at any safe threshold.

Just how far are we from preindustrial levels of greenhouse gases? Prior to the Industrial Revolution, greenhouse gas concentrations in the atmosphere were fairly stable. The concentration of carbon dioxide was about 280 parts per million (ppm), compared to today's value of ~425 ppm. Natural carbon sources, including respiring plants and animals, volcanoes, and wildfires, were balanced by carbon uptake from the oceans, plant photosynthesis, rock weathering, and more.

Land clearing and fossil fuel burning began thousands of years ago but surged with the Industrial Revolution. Greenhouse gas emissions skyrocketed as the global population grew and energy use per person rose. We are still setting yearly records for greenhouse gas emissions today—roughly 370 million tons of methane pollution from human activities each year and 38 billion tons of fossil carbon dioxide pollution globally in 2023 alone.

In class, students inevitably ask me, "When did people first link greenhouse gases to global warming?" I call this period of awareness the *Cognocene*. The physics behind how greenhouse gases warm the Earth has been clear for centuries. Scientific giants such as Jean Baptiste Fourier, Svante Arrhenius, Eunice Foote, John Tyndall, and P. C. Chamberlin—Nobel Prize winners and beyond—probed the greenhouse effect through careful experiments, theories, and models beginning in the 1820s and throughout the nineteenth and twentieth centuries. Eunice Foote filled glass cylinders with different gases in the 1850s and examined the temperature of the gases in sunlight. Referring to carbon dioxide, which yielded the highest temperature, Foote wrote, "An atmosphere of that gas would give to our earth a high temperature; and if as some suppose, at one period of its history the air had mixed with it a larger proportion than at present, an increased temperature from its own action as well as from increased weight must have necessarily resulted."

As early as the 1890s, Arrhenius even predicted a rate of warming that proved to be remarkably accurate (in °C): "Any doubling of the percentage of carbon dioxide in the air would raise the temperature of the earth's surface by 4°; and if the carbon dioxide were increased fourfold, the temperature would rise by 8°."

Through 2022, the Earth had already warmed 1.2°C through a CO_2 increase of 50 percent and a methane increase of 160 percent. Recent work suggests a doubling of atmospheric CO_2 will increase average global surface temperatures by 1.5 to 4.5°C compared to preindustrial conditions. Arrhenius's estimate from over a century ago still falls well within today's best estimate. His projection is also consistent with Exxon's internal estimate from their models of warming from the early 1980s for a doubling of CO_2.

If the CO_2 emissions from burning natural gas use weren't enough to spur climate action, methane leakage should be—because natural gas is nearly 100 percent methane. Some of that methane leaks out all along the supply chain—from oil and gas wells to the pipelines under our streets to our homes and buildings. There, gas stoves and other appliances leak unburned methane and emit pollutants into the air we breathe when gas is burned.

Our homes are therefore a fitting place to begin reducing both carbon dioxide emissions and methane leaks. Gas appliances, including stoves, water heaters, and furnaces, emit one-ninth of U.S. carbon dioxide emissions when lit. More than 60 percent of U.S. households use gas for heating or for cooking, and nearly fifty million U.S. households have gas stoves.

Home appliances are also the least studied part of the natural gas supply chain for methane emissions. My group published some of the first complete measurements of methane emissions from gas stoves and water heaters. We are still sniffing out surprises as we sample homes today.

My students and I park on a hilly Oakland street in spring, where an apricot tree hides the left side of a 1940s-era house. It's two stories, with leaded glass windows and cream horizontal siding. I open the car door into an apple tree pruned carefully to fit in the narrow rectangle between the driveway and sidewalk to the front door.

Homeowner and fervent gardener Kathy Simon bounds down the driveway to greet us. The spring in her step echoes the tight coils in her dark shoulder-length hair. Simon teaches communication—"helping people collaborate and navigate conflict" she tells me—with "people" referring widely to couples, parents, teachers, and administrators. Based on her experience and graduate research, Simon wrote *Moral Questions in the Classroom*, a book that examines how high schools can help students better explore issues that they care about and that are essential for democratic citizenship. She raises questions such as how to balance individual freedoms with the needs of society, how the interests of the public align with or differ from those of corporations, and how citizens can influence policy issues they care about.

Simon is also an early adopter of green technologies. "I've been interested in reducing my carbon footprint for a long time," she said. "We bought this house in 1997 and put solar on it around 2005. I don't remember if there were incentives at the time. Not many people were doing it back then, but I thought this is probably the single biggest thing we could do to reduce our carbon impact. So we did it."

She turned next to her vehicle to extinguish another big source of pollution in her life. "We bought a Prius in 2004, which was also early in the curve," and replaced it with a Chevy Bolt EV in 2019. "I had a hybrid for many years but eventually felt I no longer wanted to run both of those technologies—gasoline and electric—side-by-side in a car."

Now Simon's contemplating "electrifying" her home—getting rid of all fossil power and gas appliances. That's one reason she answered my group's request for volunteers to measure methane emissions in their homes—to understand how much pollution her appliances emit.

Before starting our measurements, PhD student Yannai Kashtan and I enter the cellar crawl space by Simon's garage to examine her gas water heater and furnace. We immediately smell a gas leak. Using a handheld methane wand, we trace it to a fitting on her water heater. We aren't surprised to find a leak because our previous research has shown that storage-tank water heaters, like Simon's, emit most of their methane while off. She will need to call her utility or repairman right away.

We lug our fans and other gear upstairs into Simon's home to measure methane emissions from her stove. We unspool two white Teflon sampling tubes and snake them through her house. One draws air from her kitchen back to our analyzers. We suspend the tube's end shoulder-height on a stand in the center of the kitchen to sample the air before and after lighting her stove. We unroll a second tube down the hall into a bedroom to test how far pollution from gas flames travels.

We began this set of studies a few years ago, when we realized the EPA was not including methane leakage from gas appliances in their estimates of U.S. greenhouse gas emissions. Our measurements include how much methane leaks into the air when gas appliances are off, how much unburned methane puffs into the air each time gas appliances switch on and off, and how much methane goes unburned while appliances are lit. We also measure the carbon dioxide pollution that comes with burning gas as a fuel.

Today, we begin by measuring the volume of Simon's kitchen and the background gas concentrations in her air. In almost all cases, including this one, carbon dioxide concentrations remain constant until we light a stove burner or oven, after which CO_2 levels rise.

But methane acts differently. The concentration of methane in Simon's kitchen climbs slowly but steadily even when her stove is off. We almost always find an increase in methane concentrations attributable to leaks in kitchens with gas stoves. In our first gas stove study, for instance, we found that three-quarters of all methane emissions occurred while the stoves were off. We estimated that this steady bleed of climate-busting methane from gas stoves in U.S. homes equaled the yearly emissions of half a million U.S. cars—not including the additional methane that leaked while extracting and delivering the gas to homes or the carbon dioxide pollution arising from burning gas as a fuel.

Simon watches methane concentrations rise on our analyzer's display while her stove is off. They rise even faster when her stove is lit and when methane puffs into the air in the delay between turning the burner knob and the flame lighting or extinguishing. Seeing greenhouse gas emissions in real time is enough to motivate her change. "Once I saw what was happening," Simon said later, "I knew I had to do something to reduce my emissions."

But watching greenhouse gas emissions in real time wasn't the sole motivation for Simon wanting to change. We also measured other pollutants generated in the flames of her stove: cancer-causing benzene, carbon monoxide, and NO_x gases such as asthma-triggering nitrogen dioxide (NO_2). We had found these same pollutants in dozens of other homes we visited, too.

Most people's exposure to benzene comes from breathing first- and secondhand cigarette smoke and from gasoline-powered vehicles. On average, the gasoline sold in the United States contains 1 percent benzene. Until my group started our work, no one had quantified whether—and how much—benzene forms when gas stoves are lit, or how high benzene concentrations reach indoors as a result. Benzene is known to form outdoors in open flames—in refineries, for instance, and in the flares of oil and gas fields.

Simon watches as our AROMA benzene analyzer dings, marking another fifteen-minute measurement cycle. "Look!" says technician Colin Finnegan when the first benzene readings pop up as blue dots on the screen. Measuring Simon's 2003 Hotpoint stove, we are finding some of the highest benzene emissions we have ever recorded. After turning Simon's ventilation hood on and setting her oven to 475°F in a "bread-baking" scenario, we watch the benzene in her kitchen air jump a hundredfold in less than an hour, from a background of near 0 to 11 parts per billion (ppb) benzene. According to both the World Health Organization (WHO) and the state of California, exposure to 8 parts per billion of benzene for more than one hour is hazardous. What we found easily exceeded that value for hours while her stove was on and after it was off. WHO also concludes that for cancer effects "there is no known exposure threshold for the risks of benzene exposure," meaning that all additional benzene is harmful, and that "it is expedient to reduce indoor exposure levels to as low as possible."

We had detected benzene emissions in many other homes before, but never had we seen dangerous benzene levels appear so quickly in kitchen air.

Measuring pollutants presents a dilemma. We tell homeowners what we find, but we're reluctant to pass judgments or create anxiety about health and safety. The risks of gas stoves depend on people's exposure, how—and how often—residents cook with gas, whether they use their hood to improve ventilation (surveys show most people don't), what type of hood they have, and how well-maintained it is.

No matter what we find, we tell people to use their ventilation hood every time they cook. However, as we've begun to test how hoods perform in homes under real-world conditions, we have found that many do little to reduce pollutant concentrations. A surprising number don't even vent pollution to the outdoors. Rather, such "recirculating hoods" mix the air back into the kitchen using a fan. Outdoor vent hoods are rarely mandated by code—new construction codes should require them—and retrofitting a home or apartment with one can be expensive.

Simon's kitchen wasn't the only room where we were finding high benzene concentrations, even with her hood on. Down the hall, in the bedroom farthest from the kitchen, benzene concentrations were reaching 7 or 8 ppb an hour or two after we turned the oven on. Those values were lower than the shocking 11 ppb kitchen readings we found—you'd expect lower concentrations as pollution dilutes through home air—but still higher than anyone should breathe daily. Furthermore, concentrations of benzene in the bedroom lingered for hours above the 1 to 2 ppb chronic exposure limit set by France, India, South Korea, the European Union, the state of California, and other jurisdictions. The only thing worse than measuring high pollutant concentrations in kitchens—where gas burns—is finding them down the hall where your children sleep.

Simon's misfortune extended beyond high benzene readings to elevated levels of nitrogen dioxide (NO_2), as well. Nitrogen dioxide is a respiratory irritant that triggers asthma, coughing, wheezing, and difficulty breathing. The U.S. EPA concludes that studies "show a connection between breathing elevated short-term NO_2 concentrations, and increased visits to emergency departments and hospital admissions for respiratory issues, especially asthma."

We measured nitrogen dioxide readings of 200 to 300 ppb in Simon's kitchen for over an hour with her oven set to 350°F (rising from a background reading of only a few ppb). NO_2 concentrations also rose to 190 ppb down the hall in the bedroom and remained above 100 ppb for hours. The U.S. EPA's national ambient air quality standard (NAAQS) for nitrogen dioxide outdoors is only 100 ppb. Other countries and regions have even stricter standards. Canada, for instance, has a maximum residential exposure limit of 90 ppb over a one-hour period and 11 ppb over twenty-four hours. There's no indoor standard for NO_2 in the United States, but you certainly don't want to breathe higher concentrations indoors every day than are recommended for you outdoors.

Doubt spreads like rot when I weigh how harmful benzene and nitrogen dioxide can be. My youngest son had asthma as a child. And yes, we had a gas stove. I'll never know if pollution from gas flames caused or worsened his asthma. You can never prove that a particular person—my child or yours—gets asthma, or leukemia for that matter, from pollution. You can only show statistically that more people suffer from asthma in homes with gas stoves. A 2013 summary of forty-one studies from Europe, North America, Asia, Australia, and New Zealand determined that "children living in a home with gas cooking have a 42% increased risk of having current asthma, a 24% increased risk of lifetime asthma and an overall 32% increased risk of having current and lifetime asthma." My group's work showed that breathing NO_2 pollution from gas stoves causes 200,000 cases of childhood asthma each year and as many as 19,000 U.S. deaths.

The gas industry knew the risks of indoor pollution more than a century ago. In 1907, the second annual meeting of the Natural Gas Association of America featured the following exchange as documented in the association's proceedings. The text discusses whether gas stoves should be allowed in homes without the direct "flue connection" required for furnaces and water heaters in homes today:

Association president Kerr Murray Mitchell: "Yesterday we were discussing the matter of flueless gas stoves, and it was moved that this association condemn the use of any appliance without a flue connection. We might settle this question now."

Mr. J. F. Owens: "I move that it be the sense of the association that a flue connection for every furnace or gas stove is essential."

[And after a page or two of discussion on the risks of indoor air pollution . . .]

Mr. J. S. McDowell: "I, therefore, move that the resolution which has been offered be amended to read as follows;

"**Resolved:** That it is the sense of this association that we condemn any appliance installed in such a manner as to permit the products of combustion to enter the room."

The resolution was unanimously passed.

And yet that is exactly what we do today in nearly fifty million U.S. homes and hundreds of millions more worldwide. There are no "flues" for stoves as required for furnaces and water heaters. Rather, the "flue" is a voluntary ventilation hood that surveys show most people don't use, that often does not send pollution outdoors, and that our tests and those of other scientists show often isn't effective.

I hear the concern in Kathy Simon's voice as we discuss her results. "I knew about the greenhouse gas pollution associated with gas appliances," she said, "but I had no idea about the NO_x and certainly not the benzene pollution." She paused. "My wife passed away eight years ago of ovarian cancer. I am not making any causal link at all, but the notion of having a carcinogen in the middle of my kitchen every single day is not okay with me."

What does it take to go gas-free in our homes? Simon did it, and so did we.

Inspired by Simon and others, my wife and I decided to replace all of our gas appliances, too. Because my lab developed all of our methods for measuring gas-stove pollution in our home, my (very tolerant) wife and

I spent weeks watching our kitchen nitrogen dioxide and benzene readings shoot up every time we lit our gas stove. That pollution motivated us to change every bit as much as measuring our stove's persistent methane leaks and carbon dioxide emissions.

We first replaced our on-demand gas water heater for an electric heat pump model. We chose a $SANCO_2$ high-efficiency model that—as the name suggests—uses carbon dioxide as a refrigerant rather than the common hydrofluorocarbon refrigerants that are super greenhouse gases when they leak. We replaced our gas stove with an electric induction model that cooks faster, gives us better temperature control, and emits no NO_2 or benzene pollution. Plugging it in was easy but did require us to install a 240-volt electric line to the kitchen. No more fossil carbon pollution, no more methane leakage, and, best of all, no more needless nitrogen dioxide, carbon monoxide, and benzene pollution for my family to breathe. We replaced our gas furnace with an electric heat pump that not only heats our home carbon-free (coupled to our carbon-free electricity) but also provides summer cooling in the absence of air-conditioning.

I acknowledge that not everyone can make such a transition quickly. Many renters do not own their appliances. People in lower-income neighborhoods may need financial help to replace their appliances. The recent U.S. High-Efficiency Electric Home Rebate Act includes up to $14,000 in rebates for low- and middle-income families to make their homes more energy efficient and to help them install electric water heaters, heat pumps, stoves, and clothes dryers.

Kathy Simon reached the same decision that we did. "My stove worked fine," she said, "and I don't think I would have been inspired to get rid of it except for wanting to electrify. Learning about the pollutants from gas absolutely impacted me and fueled my desire to replace my appliances. I'm grateful for your research, because if people aren't worried about the world's climate, they're certainly worried about their health."

The surprise for Simon was how simple the transition was. She said,

"When I first thought about electrifying, I thought I was downgrading, but for a good cause. But each of my new electric appliances—the induction stove, heat pump hot water heater, heat pump furnace, not to mention my EV—all of them are just plain better than the gas versions."

Simon wrote me recently: "It happened! My utility came by last week and took out the gas meter. And I put a little baby toyon in its place. Here are the photos to prove it." One photo shows a utility worker wrenching her gas meter off the wall of her house. A new toyon plant in front of it is flagged in blue for safekeeping.

As a long-term community organizer, Simon knows that transformation starts with local action and engagement. She mentioned another Oakland resident whose home we'd sampled a few weeks earlier. "She's got one of these big beautiful old Wedgewood ranges that she loves," Simon said. "It's classic." (For the uninitiated, Wedgewoods are to gas stoves what 1960s Mustangs are for car buffs.) "She was blown away by your NOx and benzene readings, and both of her kids have asthma. She said, 'I've got to move on this.'"

Simon was hyped. "I'm on a mission now," she said. "A friend told me her gas water heater had broken and she was searching quickly for a new one. I gave her the talk about not buying gas again, and she got an electric tankless.

"Friends don't let friends buy gas, you know."

3

Planet of the Cows

After sex, food is our most personal of choices. While owning a small farm in North Carolina, my wife and I raised dairy goats. Spring kidding was a popular event with friends and families. (I once rushed home from work to catch the year's kidding, taking a previously scheduled conference call for a textbook in the goat pen for the births. After I'd hit "unmute" on my cellphone by mistake, the host of the call broke a long, awkward silence by asking, "Rob, did you just say, 'I see the placenta'?") We made cheese, yogurt, flan, and more with our goats' milk.

We weren't food self-sufficient, by any means, but we did our best with our garden and animals. We kept chickens, quail, and pheasants for eggs, feathers, and fun. I cut trees to culture shiitake and oyster mushrooms on fresh logs. Our relatively rural land was overrun with deer—so overrun that nearby Duke Forest hired professional hunters to lessen the damage from browsing and to keep the deer from starving. I took a couple of does yearly from our land because venison was leaner and cheaper than store-bought meat. Setting aside the ethics of hunting and of meat-eating in general, there are some cases where eating meat can be better for the environment than eating plants. Not many, but some.

We no longer keep goats; twice-a-day milking for six months a year ties you down. Still, the first thing we did when we bought a house at

Stanford was to build a chicken coop. We gave the coop a red front door that matches our home's. We keep four chickens for eggs: two Speckled Sussex hens and a pair of Golden Campines, a heritage Belgian breed with fancy zigzag-patterned chicks.

Scaling from one family's food habits to the needs and preferences of a billion-plus families worldwide is part of what makes climate solutions for agriculture so challenging. Food production produces a third of greenhouse gas emissions globally. Carbon dioxide emissions contribute half of those food-based emissions, mostly from clearing tropical forests for farmland and pastures. Beef production, for instance, currently drives 80 percent of Amazon deforestation.

Methane contributes another third of greenhouse gas emissions from agriculture. More than half of all methane emissions from human activities are agricultural—exceeding the combined emissions from all oil and gas wells, coal mines, and industrial activities in the world. Most of these agricultural methane emissions are from the 1.5 billion or so cows grazing the planet.

The meme of the flatulent cow has legs. An average cow emits a bathtub's worth of methane a day, which comes out to a few hundred pounds a year. The amount emitted per pound of beef or quart of milk produced varies depending on a cow's diet, health, environment, and location. Nevertheless, methane pollution and tropical deforestation meet in the fraught world of dietary choices and beef consumption.

This statement from a 2019 *New Yorker* article also has legs: "Every four pounds of beef you eat contributes to as much global warming as flying from New York to London." While the basic idea is true, how correct the statement is depends on what factors are included in the estimate and the numbers used for them—actual emissions vary based on how much methane a given cow emits and whether a hamburger eaten in the United States is viewed as a U.S. or a global product (particularly whether any emissions from tropical deforestation are included). In 2022 the United States imported 3.5 billion pounds of beef and exported

3.4 billion pounds, so knowing where your beef comes from isn't easy if you buy it at a grocery store or fast-food restaurant.

Complexities notwithstanding, cows and food production play a large, and largely underappreciated, role in global greenhouse gas emissions. My colleagues and I recently estimated that beef provides less than 1 percent of calories globally but accounts for 5 percent of greenhouse gas emissions from *all* human activities. Cattle and their kin—sheep and oxen—emit more methane than the entire fossil fuel industry. Our diets matter.

The average American eats four times more beef a year than the average person globally. If everyone in the world ate that much beef, global beef consumption would quadruple, and the three hundred million cows now slaughtered yearly for meat would jump to more than a billion. The land needed to feed them would also quadruple. But because a quarter of Earth's ice-free land is *already* used for livestock grazing, and one-third of global croplands are currently used to feed livestock, there isn't space for such an increase.

To cut carbon dioxide emissions from deforestation and to restore the atmosphere to preindustrial levels of methane, then, we have two options. Those of us in wealthier countries can either change our diets, slashing the number of cows on Earth, or change the cows' diets, using feed supplements and other metabolic methods to reduce emissions from however many million cows are left—one cow at a time. To illustrate the first option, Pat Brown, founder and former CEO of Impossible Foods, describes why he's bullish on plant-based foods replacing beef and dairy products. He is unapologetic in wanting to kill the industrial meat industry.

Brown sits before me, his bushy gray eyebrows brimming over dark-rimmed glasses. He looks more skateboarder than CEO in his loose hunter-green hoodie. The black T-shirt visible under it has a pink cow printed in a classic butcher's diagram—dotted lines denoting cuts of meat, only instead of "Sirloin" or "Rump," each in its respective place, the single word "NOPE" marks every cut, appearing dozens of times across the hot-pink bovine.

A former professor of biochemistry, Brown spent years at Stanford Medical School, where he invented the DNA microarray, a staple of modern molecular biology used to measure the expression of thousands of genes simultaneously, for which he received the American Cancer Society's Medal of Honor. As to why he left such a distinguished career to start Impossible Foods in 2009, Brown tells me: "I was not contemplating doing anything in the business world. But I just sort of challenged myself to figure out how can I have the biggest positive impact on the world, given the kinds of things I'm competent at doing. I figured it would probably have something to do with addressing global environmental issues, since I feel like that's the biggest determinant of the future of humans and our planet.

"There are 1.7 billion cows on Earth. If you calculate the total biomass of cows and compare it to the total biomass of every remaining wild terrestrial vertebrate left on Earth, the cows outweigh them by more than a factor of ten. We have literally replaced nature with cows. And those cows produce less than one-eighth of the human protein supply, so we've traded nature for cows for virtually nothing in return."

Brown isn't bashful about his goal at Impossible Foods. "There's absolutely no doubt that plant-based products will replace animal-based products over the next couple of decades; not only is the sustainability better, but the economics are vastly better because you use less crops, less labor, less fertilizer, less land, and less water. The way we achieve our mission—to turn back the clock on climate change and reverse the collapse of biodiversity, by replacing animals completely by 2035—is to build a technology platform that replaces animals categorically, as a system, for food production."

Impossible Foods' products include burgers, chicken tenders, and sausage made from plants. They can be found in Burger Kings and in more than twenty thousand other supermarkets and restaurants in the

United States, Australia, Canada, Hong Kong, New Zealand, and Singapore. Pat Brown doesn't want to convert a few consumers to plant-based foods; he wants to make a better product that replaces an industry he sees as wasteful, dangerous, and unjust.

"The thing people don't realize is that it's possible to make meat more delicious than the animal version. Cows and fish did not evolve to be delicious. People learned to like the animal version, but if you're controlling the knobs, you can take consumer feedback and make it better than the version you started with.

"We're not trying to force people to eat a plant-based product. We have to make a product that does a better job than the animal-based version of delivering what consumers value: deliciousness. When you cook meat, something happens that doesn't happen when you cook, say, green beans. Instead of it just becoming a little warmer and mushier, meat completely transforms. It explodes with aroma and flavor.

"Deliciousness was the unsolved part. Why does meat taste like meat, and why is meat from any animal immediately recognizable as meat and not broccoli? The food industry has zero curiosity; innovation is a new flavor of Froot Loops. So I knew no one had looked at this because the answer was pretty obvious: the things we call meat have way more heme."

Biochemically speaking, "heme" is related to the protein hemoglobin, which carries oxygen in our blood and is found in plants and in the blood of most other animals. Plants make far less heme than animals do. Impossible Foods' signature success was to trick plants into creating more heme than they normally would. This extra heme is why an Impossible Burger "bleeds" red like a normal hamburger.

If heme is the "how" to make a plant-based burger, I wanted more information on the "why." Pat Brown's list of reasons for eliminating meat is longer than the Chisholm Trail. Impossible Foods packaging includes the statistics: "96% less land, 87% less water, and 89% less emissions."

"First of all, let me tell you about soybeans," Brown said. "You may not know this, but the world's soybean crop contains 50 percent more total protein—higher-quality protein than beef, 50 percent more of it—than all the meat consumed globally. And the world's soybean crop is grown on just 0.8 percent of Earth's land area, so keep that in mind. People think soybeans are a horrible crop. Actually, they're an awesome crop.

"The only problem is that when we use them so inefficiently by turning them into cows and pigs and chickens, you need to grow *way* more than you should have to." U.S. farmers grow seventy-five million acres of soybeans a year, but almost three-quarters of it feeds cows and other animals.

The efficiency argument embedded in Brown's explanation is the key to understanding why eating meat requires so much more land, water, and other resources than eating plants: energy transfer between organisms is only about 10 percent efficient. If a cow converts 10 percent of its food to biomass, and a person who eats a steak or hamburger converts another 10 percent, then they've incorporated only 1 percent of the calories available from the soybeans. Cut out the middleman—or middle cow in this case—and you incorporate ten times more. This inefficiency is why Cornell ecologist David Pimentel wrote, "If all the grain currently fed to livestock in the United States were consumed directly by people, the number of people who could be fed would be nearly 800 million." Switching to a plant-based diet reduces the land needed for agriculture.

Saving water is another benefit. The packaging for Impossible Burgers claims "87% less water" compared to a hamburger. A pound of soybeans takes about two hundred gallons of water to grow. Producing a pound of beef requires about eighteen hundred gallons of water, including the water needed to grow the grains and grasses in the feed and the water the cow drinks. The amount of water embedded in a hamburger always outweighs the water used to produce a plant burger.

Irrigation really tips the balance. Brown said, "A paper that was published maybe six months to a year ago reported that more than 50 percent of the entire Colorado River basin water is used specifically for beef production. Cattle are water-intensive." I went to the literature to check the numbers, and he was right. The authors of the study wrote, "We find irrigation of cattle-feed crops to be the greatest consumer of river water in the western United States, implicating beef and dairy consumption as the leading driver of water shortages and fish imperilment in the region." The authors also determined that irrigating cattle feed crops was the single largest consumptive water use nationally in the United States and accounted for 55 percent of all water consumption in the Colorado River basin. I'm flooded with environmental statistics daily, but this one shocked me: Half of all the water consumed from the Colorado River goes to irrigate plants fed to cows?

As I write these words, Lakes Mead and Powell, the two largest reservoirs in the United States, have dropped to their lowest levels since their dams were built more than half a century ago. They're both inching toward what's ominously called a "dead pool," the level at which the water level gets so low the dams can no longer produce hydropower or deliver water. Lake Mead is being emptied by a megadrought turbocharged by climate change and by water demands from Los Angeles, Las Vegas, and . . . cows.

So there's more land required for that hamburger or steak. Check. More water. Check. And definitely, yes, more greenhouse gas emissions. Brown continued, "Historically, as cattle production grew, the biomass, on a huge fraction of Earth's land, was turned into carbon dioxide through deforestation; just look at what's happening in the Amazon right now. And if we chase the cows off the land, and allow biomass to recover, you're basically reversing that movie of the burning Amazon, and turning back the clock.

"The other thing people don't take into account is that if you stop emitting carbon dioxide, well, you're still stuck with a lot of CO_2 in the atmosphere, unless you use photosynthesis to remove it," he said. Carbon dioxide lasts for thousands of years in our air. "But if you stop emitting methane, it has a nine-year half-life, which means you immediately get net negative greenhouse gas emissions—you get decay that's not being replenished by ongoing emissions. A huge potential opportunity in getting rid of animal agriculture is that methane decays quickly, reducing total atmospheric greenhouse gases. And methane in agriculture comes mostly from cattle."

The Impossible Foods packaging reads "less land," "less water," and "less emissions." The precise numbers on emissions cuts and resources used may be debatable, but the overall conclusion isn't. "When I started looking into it," Brown said, "I quickly realized that the use of animals in food technology is, by a very large margin, the most destructive technology on Earth today, and very likely, the most destructive technology in human history." Brown calls our world "Planet of the Cows."

He added, "Any ecosystem in which there are cows looks like any other ecosystem in which there are cows, and you've basically replaced the native flora and fauna with cows and what cows eat. And 'Cowboys and Indians' isn't a myth. The cowboys drove indigenous people off their lands in the United States, and in Australia, South America, and elsewhere to create space for their cows, and it's still happening."

I asked him about how much of a difference he thinks "regenerative agriculture" can make (as seen at farmers markets)—attempts to manage land more holistically and restore soils through practices such as composting, mulching, "no-till" agriculture that eliminates destructive plowing, and planting "cover crops" that protect bare soils from weeds and erosion.

"Regenerative agriculture is an oxymoron," Brown said. "It's clean coal. Here's a thought experiment. We're going to burn down the Amazon and put cows on the land and practice regenerative agriculture. Are we better off? Ridiculous. It's just propaganda and window dressing that the industry uses based on essentially no science, and it defies common sense."

Few people, including all the supporters of regenerative agriculture I know, would argue that denuding the Amazon to raise cattle is a good idea—this comparison is unfair. Regenerative agriculture is needed to manage our agriculture more sustainably. Books like my friend Liz Carlisle's *Healing Grounds* explain why regenerating healthy soil, in particular, is needed for climate and our well-being. Slogans such as "It's not the cow, it's the how" capture the importance of best practices for how cattle and crops are raised.

Nevertheless, grazing a cow on a pasture or feeding it with crops grown more regeneratively does not negate the methane burps or resource inefficiencies Pat Brown highlights. And Amazonian deforestation is happening today primarily because of increasing demand for beef.

Brown's list of reasons to take down industrial animal production includes other factors, including animal ethics and dangerous antibiotic resistance. "People will look back and think, 'How could we have been doing this?'" he said. "Just the cruelty that we impose on other animals that have feelings. That we do this as a society makes us numb to the pain of other creatures, including people."

Eliminating large animal feedlots would also make us healthier and safer, he said. "New epidemics entering the human population have come from encounters with animals used for food, either hunting and killing them, or expanding land for animal agriculture. So you reduce that risk. You also reduce the incubation of antibiotic-resistant bacteria because half of all the antibiotics used in the world—probably more than half—are given to animals. Chronic exposure to antibiotics is the perfect way to select for resistant organisms. We're incubating them in our animal farms."

Scientists agree. Tetracyclines and other antibiotics are widely fed to beef cattle daily, especially in concentrated feedlots. Doctors writing on the subject in the *American Journal of Public Health* concluded: "antibiotic resistance in humans is promoted by the widespread use of nontherapeutic antibiotics in animals." They also wrote, "Of all antibiotics sold in the United States, approximately 80% are sold for use in animal agriculture."

When I finally asked about nutrition, Brown got a little defensive. "One of our core principles is that we'll never make a product that we don't think from a health and nutrition standpoint is better for the consumer than what it replaces. Note that it's better for the consumer *than what it replaces*, not better than anything else they could possibly eat. Because the goal here is not to replace a kale salad with an Impossible Burger, it's to replace a burger made from a cow with an Impossible Burger.

"Based on what we know about nutrition and health, ours is a better product for consumers. The first product we launched was lower in calories, lower in total fat, lower in saturated fat, and had no cholesterol. And by the way, same protein, same protein quality ('PDCAA score,' so to speak, for the aficionados), same bioavailable iron, same or higher levels of all the micronutrients that people think of meat as a good source for, and higher fiber."

I searched for independent assessments. A comparison from the Harvard Medical School concluded: "The good news: Meatless burgers are a good source of protein, vitamins, and minerals. The bad news: Meatless burgers are heavily processed and high in saturated fat." Another recent analysis, in the journal *Food Science and Human Wellness*, reached a similar conclusion: "Modern meat analogue products can offer roughly the same composition of nutrients as traditional meat products, albeit with many different ingredients and a high level of further processing."

Although nutritionists may be on the fence concerning some of the health benefits of Impossible and other plant-based burgers, they confirm the health benefits of plant-based diets generally. (One could in fact replace that hamburger with the kale salad Brown referred to.) Many studies have found that plant-based diets, especially those rich in high-quality plant foods such as whole grains, fruits, vegetables, and nuts, lower the risk of heart disease, obesity, gastrointestinal cancers, type-2 diabetes, and other maladies, particularly when compared to processed red meats. And plant-based diets do require less land and water and yield fewer greenhouse gas emissions—just as the Impossible Burger packaging states.

Reading the list of "Ingredients" on the Impossible Burger packaging led me to consider personal preferences and international markets. Brown described how Impossible Foods is tackling international tastes. They began selling food in Hong Kong in 2020, for instance. He said, "Little known fact. China is the world's number one consumer of beef. They consume more beef than the U.S. or Brazil, so the products we've developed are already well suited to their market. We're not going to synthesize a bunch of fake flavors to produce fifty kinds of dumplings for China. Instead, we provide what the animal produces—the raw pork, beef, or chicken, which chefs and people there can use to make whatever they want."

What people want includes cheese, so, not surprisingly, the dairy industry is another animal production industry that Brown seeks to overturn. Cheese consumption has doubled in the United States over recent decades and now outweighs milk consumption. Brown cofounded an earlier company called Kite Hill that uses almond milk to produce yogurts, cream cheese, ricotta, pastas, and dips that are sold in Safeway, Whole Foods, Amazon stores, and elsewhere. He said, "When I first founded Impossible Foods, we started working on cheese as well as meat products, and we spun off a company that makes plant-based cheese and yogurt and various things out of almond milk. But you're never going to take down the dairy industry with almond milk."

Beyond all the data and rationales for promoting plant-based foods, Pat Brown ended our discussion with his personal motivations. "This is a blast," he said. "I have fun with Impossible Foods. I love making discoveries and inventing things, and I was highly motivated by trying to do—it sounds cheesy—good things for the world. I'm super-confident, and I hope, obviously, that a thousand other solutions are being worked on at the same time."

Given his confidence, I asked if he had thoughts or advice for young people concerned about the climate crisis we're facing. "You should be concerned," he said, "but instead of wringing your hands, step up and do something about it. Despite what Elon Musk would like you to believe, this is, by any measurable factor, the best frigging planet in the universe. You better take care of it because you're going to have to live on it."

"I love cows, by the way," he added. "I think they're awesome creatures. I just wish there were like a millionfold fewer of them."

I needed time to ponder what we'd discussed. How realistic is it to eliminate industrial meat production—for everyone in the world to stop eating beef? And what other options or technologies will we use when everyone doesn't? Not long after my conversation with Pat Brown, I took a fishing trip with my eldest son and some college friends. Partway through the trip, we found ourselves with unexpected free time for an Impossible Burger taste test. I needed first to find a grocery store that carried the brand in the conservative logging country of rural Oregon where we were stalking steelhead. But before shopping, I stopped along the North Umpqua River, at a forest that resembled charred asparagus stalks tagged with QR codes.

I bend over to scan QR codes—black-on-white-on-black—charred black. I chip the bark away at the base of one of several burned Douglas firs each marked with a QR code and a six-number tag—002762 or 065029, for instance—and find the cambium dry and damaged. These are "hazard trees" marked for removal because they stand near Oregon's Highway 138 and could fall across the road, endangering motorists. I never expected the Umpqua National Forest to be inventoried like a K-Mart aisle.

Blue dots also mark legions of scorched trees between Glide and Steamboat, Oregon, like remnants of a twenty-mile paintball war. Each dot marks a tree for salvage logging; these trees were compro mised in the Archie Creek Fire in the fall of 2020. Oregon fires that year burned more than a million acres and destroyed thousands of homes. Oregon's misery was only part of a record fire season across the western United States and Canada that caused almost $20 billion in damages while torching more than ten million acres and ten thousand buildings.

I'm here at the tail end of another climate-related event, the sweltering heat wave of June 2021. Earlier this week—on June 28—Portland shattered its all-time temperature record at 116°F (47 °C). Four hundred miles to the north, the town of Lytton, British Columbia, set the highest temperature ever recorded in Canada at 121°F (almost 50°C). Lytton burned to the ground the next day.

Climate change intrudes. I didn't come to Oregon to study a record heat wave or charred trees but to escape climate change research. I spent a few mornings on the North Umpqua River with a fishing savant and Bend, Oregon, resident who grew up here and preferred to remain anonymous. He lost part of his business in the previous summer's Archie Creek Fire. He and the other locals I spoke with are exhausted from uncertainty, destruction, and loss.

Unprompted, he told the story of a high school friend whose father was an incident commander, the person responsible for every part of emergency responses to a fire, including firefighter safety. "My friend said his father came home from a particularly bad fire and promptly retired early." When asked why, his friend's father said, "'These fires are different. They aren't what we trained for. They create their own weather patterns.'"

I remembered his words the next week when the Bootleg Fire started southeast of us and charred four hundred thousand acres of southern Oregon. Marcus Kauffman, with Oregon's Department of Forestry, told CNN the fire is "so large and generating so much energy and extreme heat that it's changing the weather. Normally, the weather predicts what the fire will do. In this case, the fire is predicting what the weather will do." Extremely hot fires like Bootleg generate towering clouds that create their own thunderstorms, winds, and lightning. Kauffman added that "dense fuels are extremely dry from a prolonged drought" and from weather that has been "consistently hot, dry with near single-digit humidity." Ninety percent of Oregon was under what the U.S. Drought Monitor defined as exceptional, extreme, or severe drought conditions.

Everyone I know in the western United States or British Columbia has a recent fire story. One friend barely made it out of her home in July of 2018 when the Holiday Hill Fire burned the Fairview/Cuesta Verde neighborhood of Goleta, California. Only when she saw flames from her bedroom window did she grab her purse and dogs and flee. The 9:00 p.m. fire that destroyed many houses on her street would have been deadlier had it happened while she and other residents were sleeping.

Back on the North Umpqua, fire was closer than we realized. After we finished fishing early on July 5, helicopters blade-slapped above us and smoke scented the valley. The Jack Fire that would eventually burn twenty-five thousand acres had just begun a few miles upstream, past Steamboat, Oregon. We had to talk our way through a Department of Transportation roadblock on Highway 138 to collect our gear. While we discussed what to do, a yellow Sikorsky water-drop helicopter passed so close overhead that I could read its red-lettered serial number. Its water-uptake tube dangled like an umbilical cord as it prepared to suck water from the river next to us. "Discussing what to do" quickly turned into an evacuation order.

An unsettling inconvenience for us was a dangerous nightmare for locals. Like last year, our fishing host had friends with animals (dogs, horses, and more) who had to move out of the fire's uncertain path. When we offered to help, he declined politely. We threw our gear into cars and drove west to Roseburg.

With time to kill and a few unexpected nights in town, I proposed an Impossible-to-beef-burger cook-off. My son and friends humored me.

First, I had to find Impossible Burgers in the former mill town of Roseburg, rural Douglas County, Oregon. Roseburg isn't a liberal enclave like Portland or Eugene. It was once called the "Timber Capital of the Nation" for the hundreds of sawmills it hosted and the billions of board feet it produced annually at its peak. Johnny Cash immortalized Roseburg in the 1960 song "Lumberjack": "Ride this train to Roseburg, Oregon. Now there's a town for you and you talk about rough."

I couldn't find Impossible Burgers at the local Safeway, Albertson's, or Walmart (though each had a competitor's plant-based products). I finally found them at the local Fred Meyer, preshaped in hockey-puck two-packs. I put four patties in my cart with a pound of organic ground beef for comparison, along with tomatoes, lettuce, onions, cheese, and buns. We were good to grill.

The taste-test participants included two Texas carnivores—Austin-based engineers Keith Brewer and Jeff Thomas—and my twenty-something son, Rob, a mechanical engineer who had recently moved from SpaceX to a prefab housing start-up. Keith and Jeff both commented on packaging that promised less emissions and water and land used when eating an Impossible Burger patty instead of beef from a cow. I explained the rationale that eating plants is more efficient than eating animals that eat the same plants. They were fast learners (with no intention of fasting).

We grilled the beef and plant burgers separately over charcoal in the backyard of the only place we could rent on the spur of the moment, and then each of us put two buns and burgers on a plate. Platters rippled down the table like the Mexican flag: green Bibb lettuce, white cheddar cheese, and sliced red tomatoes.

It wasn't a blind taste test; we sampled burgers in the order of our choice. Keith, the most committed carnivore, started with the Impossible Burger, his first-ever plant-based burger.

"Pretty darn good," he said, breaking his long silence. "It's a good flavor and it's a burger-like flavor. It feels like a burger, too."

Jeff was a little less bullish. "If you brought this from a restaurant in a sack and gave it to me, I wouldn't say a thing. It's only in the side-by-side I can tell. I've got drips here from the meat, but not from the Impossible Burger. It's a little drier, but the difference is minor."

"The real burger *is* more moist," Keith said. "Still, if I sat down and ate this Impossible Burger, I'd be perfectly happy with it and not really know it wasn't a hamburger."

My son noted a difference in texture: "The Impossible Burger is more stable than the beef burger and better in terms of structure. The beef burger was more crumbly." Then he added "Actually, the clear winner is the caramelized onions."

Generations matter in the fight for climate. The bar for switching was lower for my son, who regularly eats plant-based burgers and sausages. "A good 15 to 20 percent of my friends are vegetarian," he said, adding "Eating beef feels 'irresponsible.' I just don't feel good about eating meat generally in terms of what it does to the environment, even though I like it a lot. It seems shortsighted. There's an animal rights angle, too, but that isn't the main issue for me."

I asked what it would take for him to switch. "I'd like it to taste better, but it doesn't have to. It just needs to be close. I'm happy to switch to veggie burgers because I'm conflicted about the impact of beef on the environment."

I'm with my son. Truthfully, I crave neither beef burgers nor Impossible Burgers. Fish is the only meat I eat regularly anymore—a once-a-week meal from a local community-supported fishery, H&H Fresh Fish.

Eating hamburgers less often will help save water, emissions, and land, but it won't zero out methane emissions from cattle. If Pat Brown doesn't accomplish his goal of erasing the industrial meat industry entirely, how can we reduce emissions from however many million cows remain?

Beyond changing *our* diets, we'll need to change theirs.

In 2020, Burger King launched a "reduced-methane Whopper" in cities such as Los Angeles, New York, Austin, Miami, and Portland. Their claim: "We've discovered that feeding cows with relatively small amounts of lemongrass, during the three-to-four-month 'fattening' stages of production, reduces their methane emissions by up to 33% on average. Adding 100 grams of dried lemongrass leaves to the cows' daily diet makes a significant difference."

They launched the campaign with a regrettable, surrealist video starring Mason Ramsey, the "Walmart yodeling kid." Dressed in an all-white suit, Ramsey busts through swinging doors in a cow's rear singing "When cows fart, and burp, and splatter, well it ain't no laughing matter. They're releasing methane every time they do." (For once, the YouTube comments at the bottom were almost kind: "They did our boy dirty with this one"; "Being in this commercial would get you bullied for life"; and "Dear Mason, you shouldn't say yes to every offer!")

Problems arose. The ad campaign buried in fine print that the one-third value applied only for the brief period cows were fed lemongrass in a feedlot environment. Other scientists estimated that even if the lemongrass worked as advertised, the reduction in total emissions over the life of the cow would only be about 3 percent, not 33 percent, because most cows are only in feedlots briefly at the end of their lives.

Can feed supplements really cut methane emissions from cows? And how would you scale feed supplements to billions of cows? To understand the likelihood of cutting emissions cow by cow, I visited researchers (and cows) at UC Davis's CLEAR center (short for Clarity and Leadership for Environmental Awareness and Research).

There, director, professor, and extension specialist Frank Mitloehner and his professor colleague Ermias Kebreab lead global research on methane-cutting feed additives. Dr. Kebreab, who also directs UC Davis's World Food Center, published some of the first studies evaluating whether seaweed food supplements could reduce methane emissions from dairy cows.

Ermias Kebreab suggested we meet in the parking lot by his office before heading to the cattle station to see their experiments. Regarding meeting logistics, he emailed, "I am a tall Black guy with short hair so you can't miss me. ☺" Kebreab is indeed tall and moves like the runner and salsa dancer he is. He was raised in Asmara, the capital of Eritrea in East Africa. His upbringing still shapes his views on animals and nutrition today: "My uncle kept animals, and so in summer I used to go to his farm and help him take care of them. But in the academic year, I was back in the city.

"Coming from East Africa, I never took food for granted. Meat was expensive, and we got lamb or chicken at holidays and beef maybe once a week or so. The rest of the time our diet was plant-based. Good quality food, nutritious food is not just a given, you have to work for it.

"That's why my motivation has always been to make sure that people—as many people as possible around the world—have access to not just food, but nutritious food they enjoy that is culturally relevant to them. I work in Ethiopia. In the Sekota area of Ethiopia, the minister for livestock there told me that about 80 percent of children have some sort of stunting, because they have no access to animal-source food. They have access to crops and can get lots of corn and wheat. All the cereals are available, but they don't get the micronutrients required for good childhood development." Vitamin B_{12} is one essential nutrient found more often in animal-based foods.

In Africa, meat animals are more than "just" food, Kebreab adds: "Animals are your investment, your bank. If kids are sick or something, people sell a goat. You don't keep livestock just for food, they are part of our culture. When I graduated college, for example, my uncle sold a cow so that I could wear a suit to graduation—the first suit I'd ever worn. It has cultural significance, and it's not just eliminating the animal for food. A lot of people here in the United States treat animals as something only for food."

I ask how he started studying feed additives for reducing methane pollution. Initially he studied additives to improve animal nutrition and health. "You're trying to give animals just the right amount of nutritional additives so that they don't waste or excrete too much. But I realized that the environment is the opposite side of the same coin. When you're manipulating nutrition, you're also affecting what happens in the environment.

"I came to the U.S. about twelve years ago to continue working on sustainable agriculture. I developed mathematical models to estimate the environmental impact of nitrogen and phosphorus pollution from livestock, and slowly started focusing on the environmental aspects of methane. I saw some publications on feed additives and methane emissions and shifted towards that topic. In the last five or six years, I've spent 80 to 90 percent of my time working on feed additives and methane mitigation."

It was time for us to visit the cattle yard where Kebreab and students do their feeding trials. Driving to UC Davis's west campus and the experimental feedlot, we pass a mammoth new 3,300-bed dormitory complex called The Green, designed to produce more energy than the students consume while living there, using on-site solar panels and a food-waste digester. We pass experimental cornfields, olive trees, and fruit orchards. We pass signs for buildings hosting "Vegetable Crops," "Plant Science," and "Foundation Seed" facilities and, most uniquely, the "E. L. Niño Bee Lab."

Reaching our destination, we turn left into the "UC Davis Feedlot and Swine Facility" on Hopkins Road. The "FARM CREW SHOP" at the entrance sits in a weathered beige clapboard building surrounded by grand sycamore trees, pickup trucks, and signs with cowboy epigrams:

> HONK OFF BOZOS
> LUNCH 12 TO 1

> NO FIREARMS
> BEYOND THIS
> POINT

> HIPPIES
> USE BACK DOOR
> NO EXCEPTIONS

> FARM CREW:
> IF WE CAN'T FIX IT
> IT AIN'T BROKEN

plus the best COVID-spacing sign I'd seen, pictured with a Holstein cow:

> Stay 1 COW apart
> (mooove over)

I half expect to hear Merle Haggard or Jason Aldean drifting from the shop to complement the nearby tractor sounds—John Deere, of course.

Kebreab and I park near the towering Animal Science feed mill that mixes the cows' food: rolled corn, wheat straw, alfalfa hay, and leftover distillers' grains. Hoop houses at the site measure ammonia and other gases that cattle generate from their waste. In the central alley of a covered feedlot, we enter the space where Kebreab performs experiments on beef cattle.

Forty-eight blue hampers are lined up on the left like slot machines, each with a control system that measures how much food a cow eats (its "roughage intake"). Each cow is microchipped with an electronic ID tag that opens only one of the fifty or so hampers for each cow in a feeding trial (and provides no chance for a cow to horn in on its neighbor's meal). Kebreab and his students know precisely how much feed every cow eats each day of every experiment.

Halfway down the row of blue hampers stands a rolling stainless-steel cart with a silver box on top. Here, at last, is the gas quantification system used to measure a cow's methane emissions. The researchers teach the cows to visit the system one at a time, dispensing "cow cookies" (alfalfa pellets) as a reward: "We train them to use it before the experiment starts. That way you don't need to measure methane by putting the cow in a chamber or forcing it to wear a mask that changes how it behaves." Instead, every three hours or so each cow sticks its head into what looks like a flared air-conditioning duct. As they nibble the alfalfa pellets, a small tube hanging down sucks in air and pumps the cow's breath to a methane analyzer.

The cow's *breath*? Finally, we tackle cow flatulence head on. The meme of the farting cow is wrong (cue the regrettable Mason Ramsey video . . .) because cows burp the methane they produce. And before anyone jumps to conclusions, feedlots stink not because of methane, which is odorless, but because of the manure and urine that arise from cramming thousands of animals together.

A cow's methane comes from its rumen, the first compartment of its stomach. The rumen is voluminous—bigger than a barrel of oil. Inside the rumen, a microbial factory decomposes the grass, corn, and feed that the cow eats. According to Kebreab, only 2 or 3 percent of a cow's methane is farted, an amount so small he doesn't bother to measure it.

I ask why cows produce methane at all. "Microbes are breaking down cellulose," he says. Cellulose is the tough, fibrous material that makes up most plant cell walls and the reason we eat grass *seeds* instead of grass leaves, which we can't digest. "Microbes help the cow convert cellulose into usable compounds that are high in energy and that the animal can absorb into its blood. As part of that process, microbes release hydrogen, which builds up in the rumen. Other microbes use this hydrogen for energy. As part of *their* metabolism, they produce methane, which the cow burps."

Our conversation moves to how feed additives work. Intuitively, you would expect additives either to shrink the population of methane-emitting microbes in a cow's rumen or to inhibit their activity—or both.

For more than a decade, scientists have tested whether small amounts of feed additives could reduce methane emissions from cows. Early studies examined a compound called 3-NOP because it inhibited the enzyme that catalyzes methane formation in microbes. Less than 0.01 percent of 3-NOP added to the cattle feed cut methane emissions by 30 percent, while increasing cow weight gain slightly and, just as importantly, not harming milk production or quality. Since then, the race has been on to find the most effective supplements to minimize methane emissions per gallon of milk produced or pound of weight gained. The compounds tested included organic acids, plant oils and extracts, seaweed, peppermint powder, and leaves of the lemongrass plant, which Burger King used.

Red algae, such as species of *Asparagopsis* and other seaweeds, accumulate a different chemical, called bromoform, to defend their tissues from microbes and herbivores (and possibly for other reasons scientists haven't discovered yet). Bromoform is similar in composition to methane but with three of methane's hydrogens replaced by bromine atoms ($CHBr_3$). Bromoform doesn't seem to kill microbes in the cow's rumen. Rather, it disrupts the enzyme that microbes use in the last step of producing methane.

In their first seaweed study, published in 2019, Kebreab and PhD student Breanna Roque showed that adding small amounts of *Asparagopsis armata* seaweed to the diet of Holstein dairy cows cut methane emissions by two-thirds. However, the cows also ate 38 percent less food, presumably because the seaweed tasted bitter. (Like kids, cows are picky eaters.)

Because the first experiment showed that seaweed additives reduced methane emissions *and* weight gain, Kebreab and Roque went back to work. They hit the jackpot with a second algal species, *Asparagopsis taxiformis*, that had more bromoform in its tissues than *A. armata* and that is also a condiment in Hawaiian poke called "limu kohu" or "supreme seaweed." In a five-month study performed in the experimental feedlot I am standing in, they added 0.25 percent seaweed for the "low" and 0.5 percent for the "high" treatments as a percentage of the cows' food by weight. They also mixed in molasses with the seaweed to mask any bitter seaweed flavor. Remarkably, methane emissions dropped 82 percent in the high treatment without reducing a cow's weight gain, and in some cases even increasing it. Methane burps represent lost energy to the cow; if the microbes could recover some of that energy, the cow could grow a little faster. Kebreab said, "This experiment was proof to me that seaweed works exceptionally well."

Kebreab had one more box to check: whether the seaweed would alter the quality or taste of the beef or cow's milk. "We did chemical analyses and taste tests for milk and meat as part of the experiment," he said. He measured bromine and bromoform concentrations in milk from the dairy cows and found the same quality and taste. He analyzed meat chemistry and taste in a larger, longer meat experiment. Chemical analyses showed that protein, fat, carbohydrate, and caloric content of the beef steaks were identical with or without the seaweed additive in a cow's diet. Moreover, 112 people taste-tested the steaks and rated them as having the same flavor, tenderness, juiciness, and overall feel.

Kebreab said, "I wasn't surprised that we didn't detect differences in the meat or milk. Bromoform is metabolized quickly in the rumen." He added, "We also keep measuring methane emissions after the last seaweed dose to see how long the emissions decline lasts. We've seen reductions in methane emissions still at day four. The effectiveness goes down, but if it drops to say 30 percent that's still pretty good. It might be really good for pasture-raised animals that you can only visit once a week. It might be possible to include feed additives with the supplements that ranchers give their range cattle weekly, even when the cows are eating grass in pastures." He also mentioned research on feed additives in time-release form that would dissolve in a cow's stomach over many days. Kebreab is upbeat about using feed additives to cut methane emissions.

"What about climate progress in general?" I asked him.

"I'm optimistic," he said, "but everybody needs to think about it and help out in some way, even if it's just by voting. Climate change affects everything.

"Urgency is what's missing," he added. "Young people see the urgency. I'm very hopeful because my students are engaged and hopeful of making a difference. They make me more optimistic."

I wanted a closer look at his experimental cows. As befits an academic dean, Kebreab sported black leather shoes, so I traipsed alone through the gates and manure toward the cattle pens. The steers in his beef experiments are Angus-Hereford crosses, a hybrid common to ranches in the western United States. They're black, "polled," or hornless, and often have white facial patterning. An all-black steer with a red "19" ear tag snorted and backed off as I approached the fence it was leaning against. (Beef steers are more skittish than dairy cows because they're handled less often.) After being cold-shouldered by red "14" and green "40," too, I decided to head home.

Working my way back to Kebreab, I retraced my path through the maze of gates and alleys. Each had an idiosyncratic mix of ear-ringing clasps, clips, and chains to open and close. The range made me consider the challenges of scaling mitigation activities overall, but especially for feed additives in a world of a billion-plus cows—with ranchers, dairy farmers, and businesses in different places, environments, and cultures. How will we ever get millions of them to adopt such solutions?

Kebreab acknowledged the importance of developing a family of products to reduce methane emissions, not to rely solely on seaweed additives. Bromoform (in seaweeds) is not the only chemical additive that reduces methane emissions. The chemical 3-NOP, for instance, has advantages over seaweed in that it can be synthesized chemically, whereas seaweeds such as *A. taxiformis* have to be grown and harvested as a crop. Just as for red seaweed, feeding cows 3-NOP increases their weight gain by recapturing some of the energy otherwise lost to methane production. Increased growth is one incentive for ranchers or dairy farmers to use such additives—sometimes an additive pays for itself even before its climate benefits are taken into account.

In fact, 3-NOP has been trademarked by the Dutch company Royal DSM as a fine, white powder called Bovaer. The day before I visited Dr. Kebreab at UC Davis, Chile and Brazil issued full regulatory approval for using Bovaer as a feed additive for ruminants, including beef and dairy cows, sheep, and goats. This approval opens the door for producers there to market reduced-methane meat and milk. A statement by DSM read, "Just a quarter teaspoon of Bovaer® per cow per day suppresses the enzyme that triggers methane production in a cow's rumen and consistently reduces enteric methane emission by approximately 30% for dairy cows and up to 45% for beef cows." According to the company, the research behind the statement included more than a decade of studies across thirteen countries and more than forty-eight peer-reviewed papers in scientific journals. Bovaer was also approved as a feed additive in the European Union in 2022.

The prognosis for approving any methane-cutting additive here in the United States is thornier. Kebreab summarized the U.S. situation this way: "Here, it's going to be more challenging because the U.S. Food and Drug Administration's classified Bovaer and all methane-reducing technologies as drugs instead of feed additives. FDA has oversight for drugs, and it means that you have to do clinical trials—basically the same process as for veterinary drugs—whereas Europe, Australia, and New Zealand treat them as feed additives. There, you don't have to go through the stringent process for human or animal drug development."

At least three paths for widespread adoption seem possible. A regulatory mandate to reduce emissions and/or a price on methane emissions would both force companies to act. A California Senate bill passed in 2016, for instance, mandates reducing methane emissions 40 percent by year 2030 (relative to 2013 levels) for methane and other "short-lived climate pollutants." It targets dairies specifically, including reducing methane emissions from cow manure but does not mandate reductions from cow burps. This bill is in the spirit of the Global Methane Pledge—announced by the United States and the EU—to reduce methane emissions by at least 30 percent by 2030 (compared with 2020 levels). A price on methane emissions would speed the process and, I believe, will ultimately be required to reach temperature stabilization at 1.5 or 2°C increases.

A third path is the widespread adoption of "carbon neutral" products by companies—and consumer willingness to pay a little more for them. New Zealand's largest dairy producer, Fonterra, already markets independently verified "carbon neutral" butter in North America and Europe and, in New Zealand, sells carbon neutral milk ("Simply Milk"). According to the company, a two-liter bottle costs $4.00 compared to the Value brand bottle at $3.38. For the extra 62 cents, consumers are contributing to carbon offsets that fund native reforestation projects in southern New Zealand, a hydroelectric power plant in India, and energy-efficient cookstoves in Bangladesh.

Feed additives for cattle are coming, though. New Zealand approved 3-NOP in August of 2023. Fonterra already signed an agreement to reduce farm-based greenhouse gas emissions in New Zealand using Bovaer as a feed additive. Bovaer's 30 percent reduction in methane emissions is a good start, but far below the 80 to 90 percent reductions observed in Ermias Kebreab's red seaweed trials.

Hawaiian start-up Blue Ocean Barns is trying to grow and market the red algae *A. taxiformis* that Kebreab studied as a seaweed supplement that reduces methane emissions by "over 80 percent." CEO and cofounder Joan Salwen worked with Kebreab on some of the cow studies discussed earlier (and Kebreab advises Blue Ocean).

The cleverly named Mootral (a British-Swiss AgriTech company) uses a corporate slogan of "Saving our climate, one cow at a time." They claim their product, which includes citrus extracts, pelleted garlic, and an inhibitor, cuts methane emissions "up to 38%," based on research by none other than Ermias Kebreab, who found a more conservative reduction of 23 percent using their product.

But even if we could reduce methane emissions 23 percent by changing cow diets, this improvement would not address the carbon dioxide emissions associated with tropical deforestation and beef cattle. I doubt Pat Brown of Impossible Foods would approve.

So, which will it be—black or white? Holstein feed additives that reduce methane emissions "one cow at a time" or the wholesale elimination of the "Planet of the Cows" sought by Pat Brown?

Brown offered his view of feed additives and other solutions to reduce methane emissions from cows one by one instead of eliminating the cattle industry lock, stock, and barrel: "Feed additives are better than nothing. But in a way, it's kind of a waste of time, because that industry is going to be dead. And obviously, they're just little Band-Aid solutions for a total disaster." Surprisingly, though, he threw me a bone at the end: "So I'd say, yeah, they're incremental, but incrementally better is still better."

4

REVved Up

'm carving a turn on the Empire Grade near Bonny Doon, California, shimmying around coastal redwoods where burls curl over the road's edge. Dan Quick is in front. He's director of communications for Zero Motorcycles, the world's leading manufacturer of electric motorcycles. I'm here surveying transportation, the largest source of greenhouse gas emissions in the United States and United Kingdom and one of the biggest worldwide.

Quick pulls ahead. He's faster than I am, and the damp forest roads we're steering are slick. Plus, there's no way I'm sliding out test-driving this sleek electric motorcycle.

Moreover, I'm slowing to take in a knee-dragging array of damage along the way. We pass fire scars and charred empty lots where homeowners are rebuilding their lives after the CZU Lightning Complex fires torched thousands of buildings and acres.

Quick and I reach the end of the grade, slow for a stop sign, and cut south on Pine Flat Road toward coastal Highway 1—the Pacific Coast Highway. We turn north briefly on reaching the ocean, then U-turn and stop on the ocean side of the road about ten miles above Santa Cruz. A squadron of pelicans flies by at eye level, left to right. Pacific waves crash below. The air is so clear that we could see the Mauna Loa carbon dioxide observatory if only the earth were flat.

I've just toured the headquarters of Zero Motorcycles in Scotts Valley, and the new Zero electric SR/S motorcycle I'm on is as royal blue as the sky. Quick's on a gray Zero SR/F. The amazing thing, beyond the scenery, is that the bikes we're riding are so silent we can talk to each other in normal voices, motorcycles and helmets still on. There's no yelling or leaning in, and no exhaust fumes to inhale.

Quick and I discuss the sobering fire damage briefly (words like "shocking" and "disturbing" understate the devastation we've just seen, especially for him, who knows many coworkers hurt by the fire) until a stranger bounds across the highway and interrupts us. He points at my SR/S and says, "This is my next bike!"

"Tony," as he calls himself, is an unabashed electrophile. His wife works for Joby Aviation, a multibillion-dollar aerial ride-sharing company, based in Santa Cruz, that just went public the week before on the New York Stock Exchange. Joby boosters called it "Tesla meets Uber in the air." And like the motorcycles we are riding today, it's fully electric and essentially silent during operation.

Tony points to a brown Porsche Cayenne parked across the street at the Davenport Roadhouse. "That Porsche goes away when I get my new electric car. It's too expensive to run. Every six months the engine light comes on and, wham, it's $2,000. Then $2,000 again a few months later. Oil change: $700. For electric cars—tires, that's it."

Tony wants us to see he's serious about electric transport. "I'm not buying any more gas crap. My diesel Porsche gets forty miles a gallon if you go sixty miles per hour on the freeway, but it's archaic compared to a Tesla. Even if you don't go really fast with electric, you get used to being so quick off the line that gasoline cars irritate you. Every gas car you drive feels like you're stuck behind a ninety-year-old driver. It's like, 'Why am I going this slow?'

"Added to that, I have forty-two solar panels on my house, and I drive for free. I make so much power, I cram the rest up PG&E's ass every day. They buy it back. I don't pay any bills. I run my air conditioner whenever I want to. The heck with it. I'll never go back."

Transportation, including travel by road, rail, air, and water, releases one-sixth or more of fossil carbon pollution globally, mostly from passenger vehicle exhaust. It also comprises 30 percent of all U.S. greenhouse gas emissions. Transportation remains one of the hardest parts of the economy to decarbonize, however. Electric batteries are still relatively expensive, and many people can't afford a Tesla, an electric Ford F-150 Lightning, or a gas-powered Porsche Cayenne, for that matter.

Another challenge is the decades it will take to turn over vehicle fleets even after low-carbon options are preferred by buyers. In the United States, the average age for cars and motorcycles is twelve years each, seven years for buses, thirteen years for commercial airplanes, and twenty-eight years for train locomotives. When Warren Buffett's BNSF Railway buys its next diesel locomotive this year, the new engine will almost certainly still be emitting carbon dioxide in 2050. In fact, BNSF proudly states, "A typical BNSF locomotive will travel up to 4.8 million miles in its lifetime—equal to about twenty trips from the earth to the moon."

Car & Driver notes, "Standard cars in this day and age are expected to keep running up to 200,000 miles," adding "while cars with electric engines are expected to last for up to 300,000 miles." That's hundreds of gasoline fill-ups over a fossil car's lifetime. And each time someone fills up at a gas station, they've spewed another four hundred pounds of carbon dioxide pollution into the air, four and a half tons of carbon pollution a year on average for each passenger vehicle in the United States. At an average price of $3.96 per gallon, American drivers spend more than $2,000 on gasoline a year per car to slowly poison themselves.

And "poison" isn't hyperbole. Beyond the climate consequences, tailpipe emissions kill fifty thousand Americans a year and four hundred thousand people worldwide ("premature deaths" in the parlance of the health industry). Those deaths are attributable primarily to pollution from ozone and small particulates ($PM_{2.5}$).

Transportation remains a substantial source of fossil fuel pollution in Europe, too, but many more people there take public transit and ride bicycles. People in Germany and the United Kingdom are three times more likely to ride mass transit regularly than in the United States. Only about 1 percent of people in the United States commute to work on bicycles—one of the lowest rates worldwide—and twenty to forty times lower than in Sweden and the Netherlands. Studies show that bicycle commuting reduces stress, makes us healthier, and, by cutting fossil fuel pollution, improves air quality.

Air quality in the United States has improved substantially since passage of the Clean Air Act in 1970. Even so, four in ten Americans still breathe dangerous air, especially pollution from cars and coal-fired power plants. More Americans die from car pollution than in car accidents.

Meanwhile, Tony's been chatting the whole time, and his enthusiasm eventually wears even me out. "Gotta go," he says, unplugging at last. He trots back across the highway to his Cayenne parked at the Roadhouse. "Nice meeting you."

"Hey, nice meeting you, man," says Dan Quick of Zero Motorcycles. "We do this all the time, so if you want a ride, let me know."

Quick watches Tony go and says, "Dude cracks me up," then adds, "But it's funny. Two times we stopped and two times someone was like, 'Heyyyy!!'" The other time was at our first stoplight in Scotts Valley, where a random guy on the crosswalk stopped, pointed, and yelled, "I love those bikes!"

In the first silence since Tony blew in, I watch for electric cars zipping by. A few pass quickly, but gasoline cars still rule the road. When a delivery truck passes belching brown air, I think of all the studies showing how living near freeways and busy roads harms our health.

Living near a busy road makes us more likely to have strokes, heart attacks, dementia, and asthma and "lifetime wheeze" in childhood and middle age. As shown in England and Wales, men and women are both likelier to die from strokes when living within two hundred meters (650 feet) of a main road. The risks of disease and death drop as people live farther from the same roads—in this specific UK study and in many more.

According to the U.S. EPA, more than forty-five million Americans live within three hundred feet of a railroad, an airport, or a highway with four or more lanes. Children and the elderly are disproportionately harmed. So are people of color, who are four times more likely to breathe the dirtiest air. The EPA concludes that "exposure to pollutants emitted from motor vehicles can cause lung and heart problems and premature death."

Right now, though, with the wind off the ocean and the sky Dodger blue, air pollution isn't top of my mind. Some climate solutions are simply better products. Unlike my gas-powered BMW, electric motorcycles have no noise, no heat (or hot tailpipes to burn your legs), no fumes, no vibrations, no gears, and no clutch. I wish the shift to electric vehicles were as straightforward as shifting an electric motorcycle—no more one-down, five-up gear-cranking in stop-and-go traffic while inhaling tailpipe fumes from the cars around you. And based on today's straightaways, an electric motorcycle is the fastest thing I've ever driven—car or motorcycle.

Before riding with Dan Quick, I also met Sam Paschel, the CEO of Zero Motorcycles, to discuss electric transportation and climate solutions. Paschel began by discussing how Zero markets a product that is greener for the environment, better for the rider, and cheaper to operate. "We don't put 'green and eco' out front," he said. "We lead with emotion—all the torque and acceleration with no fumes, vibration, and noise. But we still pull the emissions out and help the planet. It's like we're almost tricking kids into eating their vegetables."

E-motorcycles combine better performance with lower operating and maintenance costs—there's no oil or air filter to change, for instance. Paschel mentioned that electric is smarter from a dollars-and-sense standpoint, too.

Dan Quick made the same point parked on the Pacific Coast Highway. "We never lead domestic marketing with the cost-of-ownership savings because motorcycling here in the United States is an emotional experience. You talk to Europeans, though, they love it. Fleet managers in particular love the cost savings of electric vehicles. You've got some consumables on any vehicle: brake pads and tires. You're going to replace those anyway, and then that's it."

Fuel costs are lower, too. At an average price of 10 cents per kilowatt-hour in the United States, the Zero motorcycle I'm riding costs about a penny a mile to charge instead of nearly a dime per mile for gasoline-powered motorcycles.

The adoption of electric vehicles remains curbed by consumer concerns, though, which Paschel acknowledged: "Our buyers want to know several things in addition to cost: How fast and far can it go on a charge? And how long does charging take?"

The answers: Zero SR/S and SR/F models go up to 124 miles per hour while delivering 140 foot-pounds of torque with a 200-mile extended range. Current E-motorcycles get 50 to 100 miles of range in an hour of charging.

"Sure, there are people that want to hop on a motorcycle and just ride to the southern tip of Baja with no bag and not think about it," Paschel said. "That's going to take more planning with an electric vehicle than with a gas vehicle." But most of us don't drive that way, for fun or when commuting.

In fact, most people in western industrial nations don't drive motorcycles at all. Although motorcycles comprise only a small percentage of registered vehicles in the United States, Canada, and the United Kingdom, they dominate transportation around the world. Most households own at least one motorcycle or motorbike in India, China, Indonesia, and Thailand (where nearly 90 percent of all households do). Those countries hold more than three billion people. There—and elsewhere—the most immediate benefits of electrification are cleaner air and improved health today, with additional climate benefits today and tomorrow.

Paschel highlighted the benefits for pollution and health. "You're getting a dirty vehicle off the road and replacing it with a zero-emissions vehicle. Gas-powered motorcycles are gross polluters. They're really, really bad. They're fuel efficient from a weight standpoint, and they get you from place to place, but a typical motorcycle is thirty times more polluting than a car," he said, referring to emissions of hazardous air pollutants such as particulates, NO_x, and carbon monoxide (beyond the greenhouse gas emissions).

In 2011, seeing a trend of car owners "trading in their cars and driving motorcycles instead because they believe that's the more environmentally friendly choice," Adam Savage and Jamie Hyneman of the TV show *MythBusters* did tests of their own. They drove a twenty-mile freeway and city course with three popular car models and three kinds of motorcycles, all fitted with emissions-testing equipment. Not surprisingly, the motorcycles used a third less fuel than a typical car and, hence, emitted a third less carbon dioxide. However, the motorcycles spewed four times more hydrocarbons, thirty times more NO_x, and eighty times more carbon monoxide pollution than the cars.

"In the automobile industry, governments have done a good job of forcing higher and higher standards on air pollution over recent decades," Paschel said. "But because of the low volume of bikes, the motorcycle industry has been mostly ignored—to the point where each motorcycle matters more and more." Indeed, when I moved to California I was surprised to find I didn't even need a smog check to register my motorcycle.

Beyond cleaner air, early adoption of climate solutions creates jobs. Denmark's early commitment to wind power is one reason a Danish company, Vestas, led wind turbine sales, installation, and service for years, and remains a world leader with tens of thousands of employees. Zero's four hundred full-time employees pale in comparison to Tesla's hundred-thousand-plus, but early investments in climate solutions compound to good-paying manufacturing jobs.

I asked Paschel if Tesla's success helps EV manufacturers broadly.

"Definitely," he said. "The more success Tesla sees, the easier our job is. They've put electrification and electric vehicles front and center for consumers everywhere. And I'd say the same thing about electric bicycle makers. The more electric options there are in our lives, all the way down to Bird and Lime scooters, the clearer this type of convenience becomes to consumers—the cleanliness, ease of use, and reduced maintenance."

Back on the Pacific Coast Highway, a car honks as it passes us. I check my mirrors and pull away from the ocean turnout with Dan Quick. After heading south to Santa Cruz, we turn north to Scotts Valley. On reaching Zero headquarters, I turn in my bike, reluctantly.

Driving home, and back in the world of cars, I watch a Bolt (Chevy), a Leaf (Nissan), and multiple Teslas zip by. Although most cars—even here in California—are still powered by gas and diesel engines, EVs are surging. In 2012, consumers purchased 130,000 plug-in EVs globally. Sales of all EVs reached fourteen million in 2023, more than one in six of all cars sold globally.

Who's capturing this growth? The car companies that sold the most EVs in 2023 included China's BYD, Tesla, Volkswagen, and Geely-Volvo, each of which sold at least half a million vehicles. Selling nearly two million EVs in 2023, Tesla's market value of $800 billion made it the most valuable car company in the world three times over. Its Model Y also became the world's top-selling car.

Although Tesla garners most of the press for EVs, I wanted to understand how more traditional carmakers view the EV market. Notably missing from the list above were Ford and General Motors. By staying tethered to internal-combustion engines, Ford and GM were lapped by their competitors in the fastest-growing segment of the global car market.

Playing catch-up, Ford plans to invest $22 billion in EVs through 2025, including all-electric models for the Mustang Mach-E, the E-Transit van (the best-selling van in the United States), and the F-150 Lightning, America's best-selling truck.

GM's financial commitment to EVs is even bigger—$35 billion through 2025. In announcing GM's commitment, CEO Mary Barra said: "Climate change is real, and we want to be part of the solution by putting everyone in an electric vehicle."

Barra returned to Stanford recently and spoke to our students about GM's plans to electrify its fleet. After attending Stanford's Graduate School of Business, Barra had become plant manager at GM's Detroit/Hamtramck assembly plant. "We call it Factory Zero now," she said, because it produces GM's electric vehicle platform and it reflects GM's vision of a future with "zero crashes, zero emission, and zero congestion."

To realize its all-electric vision, GM needs buyers for its vehicles. "We've been doing customer research for several years now," Barra said. "And customers have consistently said, first of all, they need about three hundred miles of range. Three hundred miles eliminates range anxiety. I think there's a lot of, 'It's three hundred, it's four hundred, it's five hundred, I can do six hundred,' but that just takes the cost up. And so I think finding the sweet spot will be something that evolves over time."

Range anxiety is quickly replaced with charging infrastructure anxiety. GM is working with start-ups and energy companies for that reason, as well as investing billions of dollars to make sure that a robust charging infrastructure is in place across the country.

Even more substantial than GM's financial investment in EVs is its aspiration to stop producing gas and diesel light vehicles entirely by 2035 and to make all new vehicles emissions-free. On LinkedIn, Barra wrote, "For General Motors, our most significant carbon impact comes from tailpipe emissions of the vehicles that we sell—in our case, it's 75 percent. That is why it is so important that we accelerate toward a future in which every vehicle we sell is a zero-emissions vehicle."

Even if the benefits vary a bit depending on where you live and how green your electricity is, EVs are vastly better for the environment.

Across Europe, every EV generates less than one-third of the lifetime greenhouse gas emissions of a comparable new gasoline-powered vehicle—including the extra emissions associated with producing EV batteries. This statistic assumes that the EV is powered by the average fuel mix for EU electricity, which is getting cleaner all the time.

The same is true in the United States. According to U.S. Department of Energy data, a mid-sized EV whose battery is powered by wind will have less total greenhouse gas emissions than a comparable-sized Toyota Corolla after only eight thousand miles of driving, a figure that takes into account the slightly higher emissions associated with manufacturing the EV. When using the average fuel mix for the U.S. electricity grid (23 percent coal and some natural gas and renewables), the break-even point is only fourteen thousand miles of driving. Even in a state with a high amount of coal-fired electricity—West Virginia uses about 90 percent coal power, the highest in the country—the electric car still generates less total greenhouse gas emissions. EVs are already better for climate and health—period. The only thing better for climate than buying an EV over a gasoline-powered car is buying no car at all.

EVs are not the only zero-emissions technology that GM and other automakers are banking on. A far smaller segment of the market today is in hydrogen vehicles. GM's Barra sees hydrogen vehicles more as a niche market that is perhaps better suited to trucking and large vehicles. "We've been working on hydrogen fuel cells for several decades now. The technology is still more expensive than the breakthroughs we've seen from an electric vehicle perspective. But we definitely see them as part of the solution, especially for larger vehicles—think Class A trucks and bigger—that have a higher load and higher weight." One reason hydrogen is often promoted in "higher weight" cases is that the heavier the vehicle (or ship or plane) is, the greater the mass and expense of batteries needed to power it. Battery requirements become prohibitive at current energy densities and costs when used to power a large truck, or ship, or airplane.

Even for lighter vehicles, not much has changed over the last decade when it comes to hydrogen cars. In 2008 the "Hydrogen Road Tour '08" featured hydrogen fuel cell cars from nine automakers, including GM, Honda, Toyota, Ford, BMW, Daimler, Nissan, and Volkswagen. The road rally was a thirteen-day cross-country trip from Portland, Maine, to Los Angeles: "31 cities in 18 states in 13 days." Because there were only sixty hydrogen refueling stations across the entire United States, tractor-trailer trucks ferried the hydrogen cars a thousand miles from Rolla, Missouri, to Albuquerque, New Mexico. (In fairness, one of the goals of the tour was to raise awareness of hydrogen technologies and the need for more charging infrastructure.)

Even today, there are fewer than one hundred public hydrogen refueling stations across the United States, and none are east of California; you couldn't drive a hydrogen car cross-country even if you wanted to. Only two hydrogen fuel cell vehicles are currently sold worldwide: the Toyota Mirai and Hyundai Nexo. They're expensive for their size ($50,000 to $70,000 in the United States) in part because not many are sold—fewer than nineteen thousand globally in 2023 and only three thousand in the United States. In contrast, automakers sold 14 million EVs globally in 2023.

As a scientist, it's not my job to pick market winners and losers, and I have to remind myself that we need diverse solutions. Still, I'm skeptical that hydrogen fuel cell vehicles will ever be cost-competitive with EVs. Hydrogen has another drawback that I'll describe below—it's the world's smallest molecule and therefore it leaks even faster than methane. When it does leak, it's ten to thirty times worse for climate than carbon dioxide, depending on the time frame used. Thus I don't believe we want to build millions of hydrogen vehicles that zip around the planet leaking hydrogen. I'm also not sure we want to blend green hydrogen into the leaky gas pipelines serving our homes and other buildings.

I do believe that green hydrogen will be important for long-haul transport and for industrial heating uses. It's also important not to generalize too much from one country, such as the United States, where only a few thousand hydrogen cars are sold annually, to the world; countries have different energy resources, investments, and needs.

Japan and South Korea are investing far more in the hydrogen economy, based largely on producing green hydrogen from renewables. In 2020, then Japanese prime minister Yoshihide Suga announced ¥2 trillion ($20 billion) in stimulus funding in response to COVID-19 to promote green projects over the following decade. Hydrogen loomed large in Japan's record stimulus package. "We will create hydrogen airplanes and hydrogen cargo ships," Suga said.

Although I've ridden in one of the fifteen thousand or so hydrogen-fueled cars currently on U.S. roads, I'd never set foot on a hydrogen-fueled ship. But my ship had just sailed in.

I hop on my bicycle one clear May morning and head to the train station. Our Golden Campine rooster mocks me by crowing as I leave. He begins crowing each spring as the days lengthen, so I train him not to by rushing outside at sunup and spraying him with a water bottle for our neighbors' sake. (Surprisingly, it works—if you wake early and persist.) Like the rooster, my wife mocks me, too, grabbing the spray bottle from under the kitchen sink and holding it out like a member of my pit crew. The rooster gets the last word this morning as I pedal down the driveway.

The Caltrain station is just a few miles from home. I catch the morning train, exit in downtown San Francisco, and walk northeast to the bay, passing Oracle Park and the giant bronze statue of Willie Mays (a few days after his ninetieth birthday). People whiz by me riding electric scooters, marked "Lime," "Scoot," or "Spin." Coit Tower rises to my left on Telegraph Hill as I continue up-bay. Just past the Pier 9 building, I pause to admire a catamaran—floating nose-in and docked beside the street—that resembles a blue-and-white manta ray breaching the water. It has two unusual masts amidships—one on each side—and a triangle stream of what look like Tibetan prayer flags suspended bow to stern. Looking closer, I notice each flag really is a flag; perhaps there's one for each country the catamaran has visited. Dark solar panels cover every square inch of its sleek surface, both curved and flat. The front left hull reads "ENERGY OBSERVER" in thick white-on-blue block letters.

I push the metal street gate open and descend the stairs to the dock, where I'm met by Katia Nicolet, resident scientist on the *Energy Observer* (*EO*). Nicolet's long, dark ponytail falls like rope down a long-sleeved blue shirt stenciled with "ENERGY OBSERVER" in white. The ship's lemniscate infinity emblem is also embroidered in white "rope" on her shirt, just above the ship's name. It and the circular green jade pendant she's wearing evoke the *EO*'s circular energy use.

The *EO* is no typical sailing ship. It's an energy laboratory that is circling the world's oceans for six years to raise awareness for climate action. It's testing renewable technologies and energy self-sufficiency in shipping—using solar, hydro, and hydrogen power to supplement traditional wind power. In fact, the *EO* is the world's first vessel to generate renewable hydrogen onboard, using electrolysis to split desalinated seawater (H_2O) into hydrogen (H_2) and oxygen (O_2).

The hydrogen made on board is stored in tanks that together provide enough power to propel and run the ship for up to eight days. This energy storage allows the crew to push through the doldrums and cloudy weather when solar power falls. In that sense the *EO* is a self-contained electric grid. "Our game on this boat," Nicolet says, "is always to balance our energy production with consumption. Most of the time, we use hybrid propulsion, both sails and electrical engines."

The catamaran we're on wasn't always the *Energy Observer*. Decades ago, it was the *ENZA New Zealand*, a racing vessel that circled the globe in under seventy-five days to win the 1994 Jules Verne Trophy. Since then, like many of us, it's put on weight. The retrofitted racing cat is now so weighed down by batteries, tanks, and desalination equipment—plus the living quarters of a five-person crew it wasn't designed to hold—that on most days it's too heavy to sail well on wind alone (hence the solar panels covering its surface that drive the propellers when needed—which turns out to be most of the time).

Nicolet and I descend a ladder from the walkway to boat level, then step onto the trampoline-like webbing at the ship's side and slip off our shoes. I duck into the open galley and commons suspended between the twin hulls. She introduces me to the captain, Marin Jarry, and chief engineer, Vincent Reynaud, who are sharing morning coffee with the crew. I accept an espresso and step toward the bow, where a giant screen displays a bird's-eye view of the catamaran covered in red rectangles denoting the ship's energy status.

"Here, the engineer can check everything," Nicolet says. "We can see each solar panel on the ship and how fast it's charging the batteries. Right now, the panels are producing eleven kilowatts of solar energy, about 40 percent of full charging capacity, because it's only ten in the morning."

Again, solar power is only part of the story on the *Energy Observer*, which also uses wind, water, and hydrogen power. Under the display's red rectangles, green bars show how full the ship's hydrogen tanks are and whether hydrogen is entering or leaving them (neither as I watch). The display also includes additional information, such as the amount of electricity the ship is consuming—shown on the screen's lower right as 3.2 kilowatts, which includes powering the espresso maker buzzing behind me like the fully caffeinated crew.

Nicolet and I finish our coffee, leave the galley seating area, and bounce up the port-side webbing to the bow. She steps onto the ship's white composite body. "You can walk on these solar panels," she tells me, pointing at the panels laminated safely into the hull below her feet. "Just don't step on those," pointing sideways to the thin plates perched between the body and the twin hulls and suspended over the seawater. "They're special—sandwiched in glass so they can collect sunlight both from above and reflected off the water below, making them a third more efficient. They're the only ones we can't walk on. Actually, the Prince of Monaco tried. We almost killed him, because for some reason he wanted to exit that way." (Before throwing the prince under the bus—or overboard, as the case may be—sovereign prince Albert II of Monaco has been a tireless advocate for renewable energy, climate solutions, and ocean conservation.)

I quickly learn not to stand on the solar panels either, not for fear of damaging them but because they're dark in the sunlight and burn the bottoms of my bare feet. I dance into the white seams between panels. "You're walking on some of the two hundred square meters of solar panels that power the *Energy Observer*," Nicolet tells me. "They produce thirty-four kilowatts of electricity at maximum performance." They're also wired to a bank of lithium-ion batteries like neurons to a brain.

When the batteries are close to full, some of the extra solar power is converted to hydrogen by the fuel cells and stored on board in the tanks. "We have one megawatt-hour of hydrogen energy storage when the tanks are full," Nicolet says. "If we wanted to store the same amount of extra energy in batteries, the boat would be fourteen tons heavier and would sink. Once we flip the fuel cells on to convert the hydrogen into electricity, the electricity goes to the batteries first and then to the propellers as we need it. The batteries are the midpoint, absorbing and redistributing all of the energy."

Hearing ratcheting to my left, I turn and see a sail rising. The white fabric is unfurling less as a traditional triangle and more like a canvas Slinky toy stretching mechanically upward. I ask about its surprising wedge shape and small size. "They work like airplane wings rather than traditional sails," Nicolet said. The *Energy Observer* uses a variable-camber "wingsail" that resembles an airplane wing standing on end. The design is more strongly curved on one side than the other to create "lift" that doubles the propulsion of a traditional sail (in the same way that air passing faster over the top of a curved airplane wing causes a low-pressure zone and—again—lift). "Best of all," Nicolet says, "they create less shade on the solar panels than traditional sails."

Beyond its special sails, the current *Energy Observer* employs another nifty trick on windy days. The ship's propellers are reversible hydro-generators; when turned "backwards," they send electricity *to* the batteries (at a loss of two to three knots of speed), instead of drawing power *from* the batteries to turn the propellers. The team monitors and selects among all of these options—solar, wind, and hydro-generation—to keep the batteries and the hydrogen tanks full, especially if cloudy weather is approaching that will limit solar power, their strongest energy source.

I asked Katia Nicolet if all of this makes her (and the crew) more aware of their own energy use. "Suddenly all of this becomes normal to you," she says, "thinking about your energy and water consumption on a daily basis, which you don't really do when you're home. We make adjustments, too. We cook warm meals mostly at lunch, using sunlight, and avoid cooking at night; we'll often have a cool meal at night to try and balance energy consumption and production. At home, you never think of how much energy it takes to make a coffee or anything else. You just pay your electricity bill at the end of the month."

What makes the *Energy Observer* unique even beyond its solar, sail, and hydro-generation capabilities is its hydrogen system. Hydrogen is an energy carrier, a way to store and transport energy long distances. Burn hydrogen (H_2) for fuel and the waste stream is pure water (H_2O)— no carbon dioxide, methane, or particulate pollution to kill millions.

That's fine, but you typically have to make hydrogen rather than mine it, so the devil in the climate details is how you make it. The cleanest "green hydrogen" comes from using renewable fuels such as solar power to drive electrolysis, the water-splitting approach used onboard the *Energy Observer* ($2H_2O + energy \rightarrow 2H_2 + O_2$). However, less than

1 percent of the industrial hydrogen manufactured today combines green power and electrolysis.

Instead, most hydrogen made today remains dirty "gray hydrogen" produced from natural gas or coal. It's generated by steam-methane reforming, an industrial approach that—as the name suggests—combines a fossil fuel with ultra-hot steam in a pressurized reactor. Nearly a hundred million tons of global hydrogen production a year generate a billion tons of carbon dioxide pollution annually, larger than the fossil carbon emissions of France and the United Kingdom combined. Most of this hydrogen is used by industry to refine petroleum, manufacture steel and other metals, and produce fertilizers and chemicals.

Just as for all climate solutions, scaling up hydrogen production will be challenging. Green hydrogen was recently three times more expensive than fossil gray hydrogen. However, the International Renewable Energy Agency projects that it could be cost competitive with gray hydrogen by 2030 as solar and wind prices continue falling and as cheaper electrolyzers grow to megawatt sizes. However, cost isn't the only barrier. As discussed briefly above, the tiny size of hydrogen gas makes it prone to leak from pipelines and infrastructure; it leaks even more than methane.

When hydrogen does escape, it isn't a greenhouse gas that traps heat directly. Rather it's an "indirect greenhouse gas," because its chemical reactions make greenhouse gases such as methane live longer. Hydrogen in the atmosphere reacts with nature's detergents that cleanse the greenhouse gases methane and ozone from air. More hydrogen means fewer detergents left in the air and a warmer Earth. Over a century, hydrogen's global warming potential is eleven times greater than that of carbon dioxide. Over the first two decades, hydrogen is thirty or forty times worse than carbon dioxide.

Beyond hydrogen's inevitable leakage, energy demands will also be a challenge for scaling green hydrogen production. One international estimate suggests that producing annual hydrogen demand from electrolysis today would require more additional power than all of the electricity currently generated in the European Union. And all that extra power and generation capacity must be carbon-free.

Every fuel has boosters and detractors. I don't see the "hydrogen economy" becoming economy-wide but rather one of a suite of options to decarbonize hard-to-electrify sectors, including chemicals production, smelting, and as we'll see in the next chapter, primary steel manufacturing (industries that require high temperatures and furnaces). As Mary Barra described earlier, hydrogen also has a place in long-haul shipping and, possibly, aviation. A Boeing 777, for instance, weighs about three hundred tons. Today's batteries are too heavy to power a large aircraft or a fully laden cargo ship across the Pacific. We need alternatives to fossil fuels that can be refueled quickly and that have relatively high energy densities. Hydrogen may fit the bill. Pound for pound, it has almost three times the energy content of gasoline. Volumetrically, though, hydrogen is a gas that's less dense than most liquid fuels and must be compressed and stored in tanks.

And like natural gas and gasoline, hydrogen is flammable and explosive (cue arresting images of the *Hindenburg* in flames). Nevertheless, most energy experts don't think that's a reason to eliminate hydrogen from our energy portfolio. Gasoline and natural gas explode, and we've learned to manage those risks well. We can do the same for hydrogen.

On board the *Energy Observer*, Katia Nicolet agrees, as we walk down the port side of the ship to the compartment with the hydrogen fuel tanks. "People think hydrogen is dangerous, but we put hydrogen in these tanks and sleep right next to them. We've been in big storms with five-meter swells, and had lightning storms and bad weather, and it's completely safe. I think hydrogen can be the missing link between renewable energies and long-term energy storage for situations where distance and weight are important."

No one on the *Energy Observer* thinks a tricked-out catamaran will replace container ships for long-haul shipping. Instead, the *En-*

ergy Observer's goal is to raise awareness of climate change and energy solutions. As Nicolet says, "Our goal is to showcase technologies. If those technologies can survive in the ocean with the salt water and the continuous movement of the ship and the storms and everything, they can survive on land."

The *EO* team takes inspiration from Bertrand Piccard, the first man to fly a hot-air balloon around the world and also the first to circle the world in a solar plane, the *Solar Impulse 2*. Piccard is an advisor to the *Energy Observer* and hails from a family of explorers—his great uncle was the first to enter the stratosphere and was part of the inspiration for Patrick Stewart's character on *Star Trek*. Describing the inspiration for his solar journey, Piccard said: "*Solar Impulse* was built not to transport passengers but to carry a message. We want to demonstrate the importance of the pioneering spirit, to encourage people to question their certainties. Our world needs new solutions to improve the quality of human life. Clean technologies and renewable energies are among them."

Looking over the compass and out the ship's front window, I see a sea of solar panels covering the *Energy Observer*, a futuristic lighthouse on the horizon, and the TransAmerica Pyramid on the San Francisco skyline. Nicolet and I discuss the future—what's needed and how she came to join the *EO*'s crew through reefs and personal grief.

"I worked on coral diseases for my PhD in Australia," she says. "In two and a half years, I watched my corals on the Great Barrier Reef die. They all bleached in the massive 2016 event. Beautiful reefs vanished in a heartbeat."

As a result, she changed careers from academic research to science outreach. "We have the technologies we need, and a lot of really motivated, smart people on the planet to do it. The question is, is there political drive to do so? I believe we can change the way we do things, the way we see energy, the way we see consumption as a whole."

The boat rocks a bit under our feet. "How?" I ask.

"We have to freak out a little bit," she says, "to be afraid, to react the same way we did with COVID. Climate change is a far bigger threat to our lives. I would love to see the same kind of reaction from the entire world about it. But stay hopeful, because if you give up, then the fight is over. . . . There will be some loss, some terrible climate disasters happening, but we can still salvage things. We can salvage, not only us, but my reefs and all the other species, too. Why do they have to die because of us? If you just let go and think, 'Ah it's already over,' then you're not going to be inspired, you're not going to be interested, you're not going come up with the next idea that might save the world."

5

Stop the Steel

lue sky peeks through the bay door at the end of a spare, metal-girded warehouse crammed with dump-truck loads of iron ore pellets soon to become steel. Along with the supports girding the warehouse I'm standing in, steel anchors our planes, trains, cars, boats—including the *Energy Observer*—buildings, homes, and home appliances. Steel companies worldwide produce almost two billion tons of steel yearly and emit three to four billion tons of carbon dioxide pollution doing so. The industry is responsible for a towering 11 percent of global carbon dioxide emissions. If steel were a country, it would be the third-biggest polluter on Earth.

Inside the warehouse, I approach two head-high piles of pellets—one gray, one rusty brown. As I sift handfuls of the differently colored piles, a one-tenth solution to climate change slips through my fingers like sand. Never before has climate action felt so tangible.

"It started as a . . . call it a crazy idea," Martin Lindqvist, CEO of Svenskt Stål AB (SSAB), told me, describing their quest to make fossil-free steel.

Diamonds are forever, except when they're drill bits. The giant drill—twenty feet long and mirroring a device in a Bond villain's lair—*griiinds* amid a flood of sparks, flames, and molten iron gushing from the bottom of the newly punctured blast furnace into the torpedo-shaped rail cars a story below us. My safety glasses keep slipping down the bridge of my nose in the heat. Wearing a hard hat and steel-toed shoes, I can't help but think warmly of my great uncle Harry, who immigrated to the United States after World War I to work "open hearth" at the U.S. Steel plant in Lorain, Ohio. He came home each day red-faced and beer-thirsty, which I'm also thinking of warmly right now.

Amid hissing hydraulics and the safety beeps announcing the drill arm's movement, Patrik Midebjörk motions me over. Midebjörk is blast furnace manager at SSAB's steel plant in Luleå, Sweden, a stone's throw below the Arctic Circle at the top of the Baltic Sea. He is wearing yellow-and-orange-striped overalls and rocks a white "SSAB" hard hat and graying goatee. As I approach gingerly, Midebjörk trips a yellow lever to open a manhole-sized trapdoor in the floor and points to the glowing iron sloshing into the "torpedo" cars below. Each torpedo holds nearly three hundred molten tons and, when full, ferries the iron across the plant for processing and slab casting. A scoreboard to our left displays a large red number that is now "62" but ticks slowly upward. "Sixty-two percent full," Midebjörk shouts to me, pointing to the railcar below. He drops the yellow lever and the trapdoor slams shut.

You often hear that climate solutions are a matter of political and personal will. While that's largely true in cases where low-carbon solutions are available, there are still industries where we lack net-zero solutions. Long-haul airline travel is one, as biofuel-based substitutes remain far more expensive than petroleum-based kerosene. Heavy industry is another, especially businesses that require furnaces for steel, cement, and aluminum manufacturing. Heavy industry is responsible for about one-fifth of global fossil carbon dioxide emissions—more than cars, trains, and planes combined. Global steel manufacturing burns more than a billion tons of metallurgical coal—more coal than the United States and European Union use altogether.

What makes steel production so carbon-intensive? First, primary steelmakers use the heat from coal to separate the iron from iron ore at high temperatures in a blast furnace (set your ovens to 3,000°F, everyone). The furnace uses coal both to melt the iron and to transform it chemically.

The blast furnace towering above me at Luleå is ringed on the bottom by dozens of super-sized vertical pipes resembling a monster car engine. The pipes "blast" high-pressure carbon monoxide (CO) made from coal through the race-car pipes into the bottom of the furnace to strip the extra oxygen found in iron ores. CO gas percolates up through the molten iron ore and extracts oxygen, forming—you guessed it—even more CO_2.

Coal thus serves two purposes in traditional steelmaking that must be replaced to make fossil-free steel: it provides heat for the ovens and the carbon used to scavenge excess oxygen from the mined iron ore.

Leaving the hisses and sparks behind, Patrik Midebjörk and I enter the silence of the control room one floor above the blast furnace. Nine flat-screen TVs track the torpedoes, temperatures, and temperaments of a surprisingly lean workforce attending the furnace and floor operations. They show me a cartoon (and real-time video) of the torpedo we stood over as it filled with molten iron. Another torpedo waits in the wings as the first one fills.

We exit the control room and—leaving the building—walk down a long ramp with orange handrails. "Hands on the rail, please!" Midebjörk says. I stop and look back at the furnace building—part Burning Man, part steampunk in its multistory glory of rusted pipes, towers, and bell funnels.

Looking back with me, Midebjörk says wistfully, "She's the grand dame. I hate it when people call her a dinosaur." Pronouns aside, I see what he means. But dinosaur she will be if Midebjörk's employer—SSAB—succeeds.

Before traveling to Luleå, I spoke with CEO Martin Lindqvist to understand SSAB's motivation for making fossil-free steel. I flew there on a plane made of steel, rode a bus made of steel on a road reinforced with steel, crossed bridges made of steel, and cooked breakfast in a skillet made of steel on a stove made of steel.

No primary manufacturer before SSAB had ever made fossil-free steel. (I'll avoid the terms "carbon-free" and "low-carbon" steel because steel is, by definition, a mixture of iron and carbon.) The task is harder for manufacturers that smelt raw iron ore in a blast furnace than for "secondary steel" makers that melt recycled scrap metal as a raw material. About three-quarters of current global steel production uses a blast furnace like the one at Luleå.

Martin Lindqvist sits, relaxed in his casual corporate attire, five hundred miles to the south of Luleå, in Stockholm. His passion for climate action is far from relaxed, however. "We have to reach the goals of the Paris Accord," he said, "and I'm personally very worried about it."

Lindqvist described SSAB's pioneering path. "Our journey to be the first company producing fossil-free steel was a long one," he said. "I think the idea came up back in 2014 or 2015. It took research, development, and a lot of money to develop our process. And to be honest, if you would have asked me a year ago if I was 100 percent sure we would succeed, I couldn't have given you that promise.

"When we started to get more and more sure about the possibilities and began talking about it externally, people thought we were—at best—optimistic. I remember especially when we were in New York, 2019, with some other steel companies. To put it politely, they thought we were naive."

I asked if SSAB's customers demanded fossil-free steel. Lindqvist said, "Our customers believe that their customers will ask for green steel in the future, and they want to be part of solving this huge problem for society. For us, this step was easy. We reached out to the Volvo Group to be one of our first strategic partners. They buy a lot of steel and are long-term customers of ours.

"Personal connections also matter," he said. "I know Martin Lundstedt, Volvo's president and CEO, very well. The president of Volvo Construction Equipment, Melker Jernberg, used to be executive vice president here at SSAB. So, it was easy to pick up the phone and ask, 'Do you want to be part of this journey?'"

I recall Lindqvist's words while crossing the Luleå plant to see his vision being realized. I arrive at a silver-cubist tower with prominent "HYBRIT" lettering down the building and "FOSSIL-FREE STEEL" in green block letters splashed across it. It's muted in greenhouse gas emissions, if not in appearance.

To produce fossil-free steel, SSAB had to replace all the coal they used to generate heat to melt the iron ore and to scavenge the ore's excess oxygen. The company settled on the idea of using green hydrogen (H_2) made from renewable energy to heat their ovens and to remove the excess oxygen. Instead of emitting carbon dioxide, both steps would yield only water—if they had a source of fossil-free hydrogen.

To get it, they teamed up with long-term business partners Vattenfall and LKAB to form a consortium called HYBRIT (Hydrogen Breakthrough Ironmaking Technology). LKAB supplies SSAB with local iron ore, while Vattenfall provides the power. "Vattenfall" is Swedish for "waterfall," reflecting the company's portfolio of hydropower (and renewable energy broadly). Vattenfall would supply the carbon-free electricity used to make green hydrogen through electrolysis ($H_2O \rightarrow H_2$ and O_2), the same water-splitting process we encountered onboard the *Energy Observer* catamaran.

I greet Åsa Bäcklin of HYBRIT, my next host at the Luleå plant. "We make the green hydrogen here," she tells me as we reach the electrolysis units. She points next to the silver tower above us, where the hydrogen contacts the iron ore to remove the oxygen. "The iron ore pellets go in on top," Bäcklin says, "and come out the bottom." I picture a cylinder packed with ball bearings flowing downward. As the balls flow down the cylinder, green hydrogen is blown into the bottom of the tower and passes upward through them.

"The green hydrogen we make is released into the shaft to contact the iron ore pellets," Bäcklin says. "It captures the excess oxygen from the ore and leaves as water vapor. You put in a ball of iron ore up there. It's still a ball when it comes out, but it's slightly smaller because you've removed 30 percent of its weight as oxygen—that's how much oxygen is in the pellets." The pellets leave with tiny holes in them (hence the industry term "sponge iron") and turn from rust-red to gray.

We walk to an open warehouse with dumpster-sized piles of separate gray and red-brown balls the size of large marbles. The rust-colored ones I pick up are from the mined iron ore that LKAB supplies. Bäcklin says LKAB calls its work affectionately "making meatballs," referring to the reddish spheres they produce. The gray balls, in contrast, started as the same reddish spheres but were converted to low-oxygen "sponge iron" using hydrogen in the tower above us.

Beginning in 2021, SSAB took batches of gray sponge iron produced by HYBRIT, melted them in an electric arc furnace fueled by renewable power, and produced the world's first fossil-free steel.

How often can you seize a solution to a tenth of climate change? I began this book with hope—and remain hopeful after meeting so many groups working on climate solutions around the world. You visit a project like Luleå and think, "It's so easy. Are we really going to fuck up the world's climate till Doomsday when so many good solutions—and people—are out there?"

We are—if we don't act faster and if companies worldwide don't have to pay for greenhouse gas pollution. And, as CEO Martin Lindqvist reminded me, nothing about SSAB's journey was "easy."

Northern Sweden provided everything SSAB needed to make fossil-free steel: renewable Vattenfall hydropower, high-quality iron ore, and local manufacturing facilities nearby to minimize transportation costs. Discussing the partnership, Lindqvist said, "We're providing a fossil-free value chain, all the way from iron ore up in the mountains down through our finished products."

In August of 2021, SSAB rolled out the first fossil-free steel made with gray HYBRIT sponge iron. Volvo produced the world's first vehicle made with it two months later. It wasn't a gas-powered Volvo S-60 sport sedan or an XC40 Recharge, Volvo's first electric vehicle. It wasn't even a Polestar 2, the new all-electric brand of cars Volvo founded. Instead, it was a sleek yellow open-load carrier for mining and quarrying ores, with "WORLD'S 1ST FOSSIL FREE STEEL" stenciled on its side next to the Volvo and SSAB corporate logos. Volvo is already producing heavy-duty electric trucks made with HYBRIT steel and plans to release more green-steel vehicles soon. As a result the HYBRIT partners are investing tens of billions of dollars to supply enough fossil-free steel to meet the growing demand.

I asked Lindqvist how much more green steel cost. He said, "Initially, we thought the cost would be 20 to 30 percent more than producing steel the normal way." That extra cost he referred to comes mainly from producing and using green hydrogen instead of coal. But there's another cost that SSAB incurs—something businesses in Europe face daily but most U.S. businesses don't—a fee to emit carbon dioxide pollution. In Europe, the polluter pays.

Two Nordic nations, Finland and Sweden, passed the world's first national carbon prices in 1990 and 1991, respectively. Sweden's carbon price rose fivefold from €25 per metric ton of CO_2 released in 1991 to €122 in 2023 (about U.S. $130 per ton). As a result, business leaders there have had to factor real carbon prices into their economic decisions for decades.

Producing green steel allows SSAB to avoid paying the fees for carbon pollution they would have incurred using coal. Referring to when SSAB was first considering the change, Lindqvist said, "At the time in Europe, too, I think the price of emitting one ton of carbon dioxide was about five euros. And the European average is roughly two tons of carbon dioxide emitted per ton of steel." A typical European steelmaker would thus have paid about €10 in carbon fees for each ton of steel they produced in the past.

But Europe's carbon price is rising—increasing the penalty for emitting carbon dioxide pollution. The EU's carbon price hit €50 per ton of CO_2 for the first time ever in May of 2021. Lindqvist discussed how such relatively high carbon prices eliminate any cost penalty for producing hydrogen-based green steel. "Now, as we speak, the cost of emitting one ton of carbon dioxide is roughly 65 €." The EU carbon price topped €100 per ton of CO_2 for the first time in February of 2023. By making its steel carbon-free, SSAB avoids paying hundreds of Euros in carbon fees for each ton of steel by eliminating two tons of CO_2 pollution. Avoiding these fees pays for the slightly higher cost to use carbon-free hydrogen in their process instead of coal. "So most of the price gap has disappeared," Lindqvist said.

What's economically feasible changes when markets price carbon pollution. Across most of the United States, carbon pollution is still free, a fact delaying the switch to fossil-free manufacturing.

Lindqvist continued: "So, I would say in the future, depending on emission rates and the cost of emitting carbon, green steel won't be more expensive at all. And in the short term when it comes to any potential premium, what are customers willing to pay? There is a premium price for this product in the beginning, but over time, I expect that green steel will be preferred. Volvo, Daimler, and other companies are already eager to buy fossil-free steel. When the rest of the steel industry moves over to this technique, it will be the new normal."

Change takes time. SSAB set a date of 2026 to convert additional facilities to green steel and 2030 as the target year to convert all of its facilities internationally. In the best of scenarios, decarbonizing heavy industries will take decades after that. I mentioned how important the age of infrastructure is, the need to turn over equipment in an orderly way—ideally at the end of its lifetime.

"Exactly," Lindqvist said. "The reason for choosing 2026 to convert our Oxelösund facility to green steel was that we needed to invest in new blast furnaces and oven batteries anyway, so we decided to invest in new technology then."

The age of infrastructure is one of the biggest challenges Lindqvist identified for the industrial world in general. "Infrastructure lasts for years," he said, "and it takes years more to build new infrastructure. In Sweden, for instance, it can take eight to twelve years to get a four-hundred-kilovolt transmission cable built. But the Swedish government is working hard to reduce lead times, which is very positive."

Groundbreaking climate work is happening in Sweden and in Nordic nations in general. In 2016, Norway committed to all new car sales being zero emissions by 2025; four of every five cars Norwegians purchased were already electric by 2022.

In contrast, companies in most other countries, including the United States, lack a national carbon price. Instead, they often use an internal (or "shadow") price for estimating carbon costs—if they employ one at all. A shadow price is a theoretical carbon price levied throughout a company to support long-term business planning and investment strategies. Shadow prices are designed to reduce fossil fuel emissions and to hedge the economic risk of a real carbon price being levied midway through a project. The median internal corporate carbon price globally a few years ago was €22 (about U.S. $25 per ton)—less than Sweden's initial price more than three decades ago.

Because Sweden and Finland have both priced carbon pollution for decades, their greenhouse gas emissions have declined by more than one-quarter since 1990. In contrast, although U.S. greenhouse gas emissions have been falling for more than a decade, they remained nearly the same in 2021 as in 1990.

Policies matter. Carbon pricing creates jobs and homegrown technologies, such as SSAB's new fossil-free steel production. Sweden produces only five or so million metric tons of steel a year, though, less than 1 percent of global steel production. China manufactures more than a billion tons yearly, half or more of the global total. According to Lindqvist, Chinese manufacturers are interested in their process, but SSAB has no current partnerships there. "We have to say no to people more or less every week," he said, "ambassadors, competitors, customers who want to come see what we are doing. There is enormous interest in our method—eventually we will spread this technique globally."

SSAB also produces secondary steel in the United States, melting scrap iron using electric arc furnaces and mixing in carbon and other additives to produce recycled steel. SSAB's facility in Montpelier, Iowa, is on track to produce recycled steel using only wind power by 2024. Wind turbines are three-quarters steel, and Lindqvist noted the symbolism of using fossil-free steel to build them. SSAB plans to convert its Montpelier facility to fossil-free primary steel production by 2026, importing HYBRIT technology to the United States and creating new jobs.

Jobs come to mind as I complete my visit to the bustling Luleå steel plant. As Patrik Midebjörk and I walk to the steelworks downstream of the blast furnace, torpedoes full of molten iron roll past us on rail lines. We step indoors and watch the torpedoes transfer their sloshing iron into giant cauldrons or ladles that slide across the shop floor. Sulfur impurities and carbon are removed from the iron here. Additives such as chromium and nickel adjust how hard, strong, and flexible the steel will be.

Once the chemistry and temperature are right, the molten steel is poured from the ladles into the slab casters we see a floor below us. Cooling water rushes past, hissing and turning to steam as it contacts the red-hot steel. Twenty-ton slabs the length of school buses roll by, heading to the transfer warehouse next door. There, each slab cools and is hand-labeled with a serial number to track its batch and date. Robotic magnets lift the cooled slabs and stack them across the warehouse floor, waiting for a train or ship (made of steel) to transport them worldwide.

Looking at the steel slabs stacked like sawmill two-by-fours, I can't help wondering about both sides of reducing emissions from steel production—what it will take for all the world's steel to be made fossil-free and how we can reduce the demand for energy-intensive steel by substituting less carbon-emitting materials such as renewable wood. (In buildings, for instance, cities are beginning to see the rise of wood "ply-scrapers.")

Both transitions will require stronger climate action and a price on fossil carbon emissions. In ending our conversation, I asked Martin Lindqvist if he (and SSAB) supported a price on emissions: "Of course we support a global price on carbon dioxide emissions. My opinion is that the process is too slow. I think it would be better if it cost even more to emit carbon dioxide in Europe and globally. When I talk to politicians, I tell them it was a good idea that they put pressure on steel companies to solve this problem."

I've spoken in Washington, D.C., at the Business Roundtable, an association of the CEOs of America's leading companies, and I attended the CEO Stakeholder Roundtable from the oil and gas industry at the 2019 climate meeting in New York City, where SSAB announced its fossil-free steel. I've spoken with the leaders of many U.S. companies who acknowledge that climate change is real and a serious problem. But I don't recall another industrial CEO telling me the price of carbon pollution needs to be higher. No caveats, no qualifiers, and no hesitation. Just a call to action.

And Lindqvist definitely calls for action. "We've shown that there are solutions to reach the goals of the Paris Accord. I mean, globally, the steel industry emits a lot of carbon dioxide. Something that was impossible a few years ago is now happening."

He mentioned additional motivation for his work. "I have three kids, including two daughters who are quite vocal about what they think needs to be done. You know Greta Thunberg is from Sweden; young people listen a lot to her. . . .

"For the first time ever, my two daughters are proud of SSAB. They thought SSAB was just another shitty company, but now that we're trying to do something that is really good for the environment, they are actually quite proud."

6.

Pipe Dreams

'm driving west on E Street in Washington, D.C., a block or two from the White House. Nathan Phillips, an ecologist and pioneer in detecting gas leaks under city streets, rides shotgun holding a folded paper road map—yes, *paper*—not because we can't use cellphones but because he's coloring the map block by block with a yellow highlighter as we track our passage across the city. Just to our east, the Yellow Line Metro runs through Chinatown past the National Portrait Gallery.

We're measuring gas leaks under city streets and sidewalks, to lower methane emissions and improve pipeline safety. Hundreds of thousands of gas leaks pepper the local pipeline networks of cities and suburbs across the United States alone, each leak a source of climate-busting methane, each leak an explosion risk. We're mapping Washington, D.C., and other cities to help get them fixed.

Our car sports more plugs than a tattoo parlor. A PVC pipe juts up from the roof holding an anemometer marking wind speed and direction continuously. A high-resolution GPS sensor pokes out like Frankenstein's monster's neck bolt to track where we are and where we've been. Air-intake tubes that dangle tentacle-like from the front bumper connect to our new methane analyzer buzzing in back.

Our work and that of many other scientists suggests the EPA underestimates urban methane emissions by a factor of at least three to five.

That gap means we aren't properly accounting for the dangers of supplying gas to our homes and buildings. In Chapter 2, we saw how methane emissions inside our homes were also much higher than EPA estimates; emissions of indoor air pollutants were similarly higher. Here, we're seeking opportunities to reduce leaks (and rare explosions arising from them) in the local "distribution" pipelines managed by public utilities that supply gas to our homes. Similar work is needed along every step of the natural gas supply chain—from oil and gas wells to pipelines to homes and buildings.

"Right on Fifteenth," Phillips tells me. We drive with the windows cracked to sniff the leaks. "Gas fart there," he says. That distinctive sulfur smell has been added to pipeline gas since 1937 when odorless methane puddled beneath the New London School in Texas and blew up, killing hundreds of schoolkids in the third-deadliest disaster in Texas history.

We've just cleared Fourteenth Street, where we drove south to mark the first half of a right-to-left U, avoiding left turns because they slow us down. We're covering the streets of Washington like a lawnmower—up and down, left and right—circling each cul-de-sac and backtracking for the diagonals. Mapping a city takes weeks.

Cooped up in a car together, we quickly learn each other's tastes in music, writers, and more. For Phillips, it's alt-rock, *Moby Dick*, veggie food (most recently—Burger King's Impossible Burgers), and the Celtics (he's from Boston, after all). For me, it's singer-songwriters such as Willie Nelson, *East of Eden*, and the Chicago Cubs and Houston Astros—legacies of my immigrant childhood.

One block later, I turn north on Fifteenth to complete the U, driving only twenty to keep our location readings as accurate as possible. Our analyzer pings.

"Bingo," Phillips says, noting the location of a natural gas leak within view of the White House lawn. He doesn't mark the leak on the map by hand. There are far too many to fit on the page, and the computer in his lap tracks those. Instead, he keeps highlighting yellow, so we know what streets we've covered. A few streets to the north our methane analyzer jackpots like a Vegas slot machine. We map half a dozen leaks on one block between I and K Streets at McPherson Square, its bronze Civil War general scowling at us from his horse.

"Cyclist on your right," Phillips warns.

Nathan Phillips is a mop-topped scientist with an infectious smile and, at times, a close-cropped beard and mustache. He cycles everywhere, faster than most people drive. I bet he sleeps in his bike helmet. Once, I went to see him in his office and couldn't figure out why he was so fidgety. Then I realized he was pedaling under his desk—pedal-powering his laptop to avoid using grid electricity. You must love someone that committed.

The slow-bleed of methane leaks we're measuring begins "upstream" at oil and gas wells and ends "downstream" at gas appliances—furnaces, water heaters, and stoves, as we saw in Chapter 2. Gas transported through large interstate pipelines is delivered to homes and buildings in the spaghetti of water mains, sewer pipes, and data lines underlying our sidewalks and streets. Some cast-iron pipes in use today date back to the Civil War generals we keep circling. They were laid in twelve-foot sections of rigatoni-like pipe with one end flared and the other end straight to slip into the next flared segment. Not surprisingly, pipes more than a century old leak. Newer distribution lines are typically plastic or protected steel to curb the corrosion that plagues older cast iron.

If our method of driving city streets is like mowing a lawn, then finding leaks is like trolling for fish. Instead of waiting for the first click of the reel when a fish strikes, though, we're tracking pings when our analyzer detects extra methane in the air over the road. When the concentration is especially high or there's a cluster of leaks, we circle the block to remeasure them. Our wind-tracking anemometer tells us which side of the road the leaks are on.

After squaring the circle to retrace our steps back to leaky Fifteenth Street, I've parked halfway up the block and put our flashers on. We hop out, grab our handheld analyzers—their red LED numbers noting methane concentrations—and use them to "sniff" curbs, sidewalks, and manholes for leaks. More often than not, we smell gas right away. These cases cluster in schools (sometimes *near* schools)—like the leak cluster we saw earlier at McPherson Square. Phillips calls such places "areas of neglect."

We use a large syringe to collect gas samples of the big leaks that we'll analyze back in the lab to distinguish fossil pipeline gas from methane generated by microbes in sewers. In almost all cases—far north of 90 percent—the leaks we find come from gas pipelines.

Over two months, Nathan Phillips and I also measured methane concentrations underground at nineteen of the highest-concentration leaks we mapped in Washington, D.C. After parking the car and finding the source of each leak, we detected explosive methane concentrations inside manholes at twelve of the nineteen locations. We phoned in the leaks to the responsible utility—receiving queries about who we were and what we were doing and assurances to have the leaks looked at. We returned six months later and found that nine were still explosive.

"This whole block needs replacing," Phillips says, as we finish collecting leaks on Fifteenth Street. We gather our materials and return to the car. To break up the mind-numbing routine, we swap places for the next hour. Phillips drives and I alternate staring at the screen and map, slowly painting my fingertips yellow from the highlighter pen.

"Keep going north on Fifteenth," I tell him after checking the map. He shuts off our car's flashers and eases from the curb. In a couple of blocks the next fish bites. As we loop Thomas Circle, the analyzer goes off another half-dozen times, with concentrations even higher than on Fifteenth Street. We don't stop, though. We'll have to return at night to sample this section of a dangerously busy road. Another Civil War general—George Henry Thomas—sits astride his bronze horse at the circle's center.

When we began the project, we expected poorer neighborhoods

would have more leaks than richer ones. Disenfranchised communities and people of color are more likely to live near highways, refineries, and petrochemical-related infrastructure. They breathe air and drink water more often laced with chemicals and pollution. They raise children more likely to have asthma and cancer. We didn't know whether similar issues of environmental justice would apply to pipeline leaks across U.S. cities.

In the cities we'd examined so far, at least, the answer was "no." Instead, we found that older neighborhoods had more leaks than newer ones. The number one predictor of a natural gas leak along streets in Boston, where we'd mapped more than three thousand leaks the year before, was cast-iron and other antiquated piping.

Why are utilities still using pipelines older than a century to move gas to homes and buildings, especially in cities of the midwestern and eastern United States and in Europe? Replacing pipelines of any kind—gas, water, sewer—is expensive. In places like Washington, D.C., and Manhattan, replacing the pipelines on a single block can cost a million dollars—disrupting lives and traffic for weeks.

Utilities, the companies that distribute gas through cities and suburbs, want to replace their older pipes. Worn-out pipes are at best a hassle and at worst an explosion risk. A customer walking her dog smells a leak and dials 911 to report it. The company sends an inspector to determine whether the leak is dangerous, especially whether it's near a home or building. Even a small leak can pool in a basement or under a house and, in rare instances, explode. Two years after we mapped Washington, D.C.'s streets, a gas-related apartment explosion killed seven people and displaced more than seventy others a few miles to the north, in Silver Spring, Maryland. The U.S. National Transportation Safety Board faulted local utility equipment for the explosion.

Leak safety more than leak size determines how quickly and *whether* a utility fixes a leak. Grade 1 leaks, as classified by utilities, found near homes or buildings are determined to be hazards to people and property and should always be repaired, no matter how small. Grade 2 leaks near a building may be deemed not hazardous now but could be soon. They're fixed by utilities most of the time. Grade 3 leaks that are far from any building and that mix into the open air might be large but never repaired because they dilute quickly in air and aren't considered dangerous, despite their smell and emissions of climate-harming methane.

We've measured some leaks repeatedly for years. We mapped a huge leak off the Massachusetts Turnpike many times in Boston. It was away from any buildings in a tangle of intersecting roads that I suspect make it too expensive and disruptive for the utility to fix. And being far from any buildings reduces its priority for repair.

On one campaign, Phillips and I and our students stopped to sample a leak. There wasn't a proper place to park, so we pulled over, put our flashers and yellow vests on, and hopped out. Halfway through our sampling a police car pulled up, siren on.

"Oh, great," I whispered to Phillips, both of us leaning over a manhole leak.

A policeman stepped out of his car and approached us. "What are you doing here? You're parked illegally."

"We're sampling gas leaks," I told him, giving him a quick rundown of our project, hoping it would spare us trouble and a ticket.

He paused, started to say something, and paused again. Eventually he said, "I'm so glad you're here. People have been calling in this leak for two years."

How can this happen in the richest nation on Earth? In truth, gas utilities would go bankrupt if they had to fix every tiny leak, and customers won't pay for them to do it. Public utility commissions (PUCs) cap how much companies can charge consumers to repair and replace pipelines in neighborhoods, and people judge PUCs by how low they keep their rates. Allocated $50 million a year for repairs by the PUC, a utility has to pay anything more than that from its own coffers. Leaks away from homes and buildings that aren't judged to be hazardous are the lowest priority and often ignored, climate consequences be damned.

We can change this. The year before our work in Washington, D.C., Phillips and I joined with our students and colleagues to publish the first public maps of natural gas leaks in any city, beginning with Boston. We mapped more than three thousand leaks across its serpentine streets, four leaks for every road mile.

When our first paper came out, with its signature map of red for roads driven and thirty-four hundred gold spikes for the leaks, people noticed. Within a day, Boston mayor Tom Menino wrote a strongly worded letter to the Department of Public of Utilities urging its chairwoman to increase scrutiny of pipeline leaks. Then Representative (and later Senator) Ed Markey wrote a letter to the federal Pipeline and Hazardous Materials and Safety Administration (PHMSA): "This study shows that we need a plan to ensure leaks from aging natural gas pipelines in Boston and other cities and communities are repaired, so that we can conserve this important natural resource, protect the consumers from paying for gas that they don't even use, and prevent emissions of greenhouse gases into the environment."

Markey was right—we pay utilities for the gas that leaks from their pipelines, ponying up for "lost and unaccounted for" gas that goes missing in our cities and neighborhoods. My experience is that the people working in utilities care deeply about consumer safety and the integrity of their pipelines. Still, when customers pay for the gas that leaks from their pipelines, utilities have zero economic incentive to make repairs.

Sometimes a scientific study catalyzes the hard work that dozens, even hundreds, of other people have undertaken for years. Labor unions in Massachusetts had called for more pipeline repairs long before our work. Consumer safety advocates pushed for greater pipeline safety. Environmental advocates promoted cleaner air with less greenhouse gas emissions and local ozone pollution.

The year after we published our Boston study, Massachusetts passed an accelerated pipeline repair and replacement bill. There were immediate benefits. Local distribution companies were allowed to spend—

and recover—more money to fix their pipelines. Massachusetts created good-paying jobs in its cities and neighborhoods. Consumers paid only a dollar or so more per household each month but also saved by paying less for leaked natural gas. The risk of rare explosions decreased, too. And, yes, the world's climate benefitted, but climate wasn't the prime motivator—public safety was. And that's okay.

Good news travels. We achieved the same outcome a year later after publishing work mapping six thousand gas leaks across Washington, D.C. Washington Gas's PROJECT*pipes* program began soon after, as a forty-year pipeline replacement program to modernize D.C.'s twelve-hundred-mile natural gas distribution system. The District's Public Service Commission approved the program and concluded it would "enhance overall system resiliency, reliability, and safety for nearly 165,000 gas customers." It would also reduce greenhouse gas emissions.

Restoring the atmosphere—returning to preindustrial levels of gases such as methane—forces us to grapple with specific dates for phasing out fossil fuels. No transition may be more complicated than drafting a plan to shut off the pipelines that supply gas to more than seventy-five million households in the United States alone.

One way to do this is to require that all new construction use a cleaner, safer alternative, beginning on some future date. Taking this approach is cheaper and less disruptive than requiring people to remove *existing appliances* and having to pay them to do so. Every new home or building plumbed with natural gas or other fossil fuel locks in greenhouse gas pollution for decades to come, and requires us to maintain expensive, leaky, and, in rare cases, dangerous and explosive pipeline networks to supply them.

Such changes are happening around the world. The state of Victoria in Australia, including Melbourne, recently passed a requirement that all new homes be powered by electricity starting in 2024. Vancouver,

Canada, has required zero-emission hot-water and space heaters in new homes since 2022. More than a hundred municipalities in the United States have adopted similar requirements, including the cities of Los Angeles, San Jose, Washington, D.C., and Denver (for commercial buildings), as well as New York State, which bans fossil fuel hookups in new homes and buildings starting in 2026 (with exceptions for hospitals and commercial restaurants). If such codes are upheld in the courts, they will change how millions of people heat their homes and cook their food.

Under lobbying pressure from gas utilities and the American Gas Association, however, and in response to these new requirements, at least twenty U.S. states have passed "preemption laws" to halt this progress, including Texas, Arizona, and Ohio.

Many questions remain: When should utilities stop laying new gas pipes under streets and sidewalks as new neighborhoods are built? How can utilities turn gas off permanently along a street or neighborhood, a process called "branch pruning" or "zonal electrification"? Must every pipeline keep running until its last customer decides to electrify? Without a specific end point for gas use it's impossible to plan a clean transition away from fossil fuels.

My local utility, Pacific Gas & Electric (PG&E), is questioning "status quo gas replacement" (their term). They have already paid homeowners in some areas to buy new electric appliances so that the utility can retire leaky gas pipelines serving their neighborhoods, avoiding the expense of pipeline replacements. For instance, PG&E has to replace twelve hundred miles of older plastic pipes installed before 1973 that fail catastrophically and have already caused several explosions for their customers.

In a project started in 2022, PG&E asked the CA Public Utilities Commission for permission to spend $17 million to electrify twelve hundred housing units on the campus of Cal State Monterey Bay. PG&E's motivation was that electrifying the residences cost less than replacing miles of brittle plastic pipes serving the customers. Retiring the gas pipelines is also consistent with Cal State's master plan to replace gas with electricity in new and existing buildings across campus and to improve indoor air quality (for the reasons we found in Chapter 2). If the plan goes through, it will be the largest electrification project to date in California.

Such projects make economic sense today and help the climate. But for an orderly transition, we need codes that mandate all future construction be electric. And while PG&E has generally supported electrification efforts, other utilities that sell only gas (and not electricity) typically haven't. For instance, the Southern California Gas Corporation (SoCal-Gas) that serves Los Angeles and Santa Barbara sued the California Energy Commission a few years ago to prevent it from carrying out policies that promote home and building electrification at the expense of gas use.

Nathan Phillips and I are back in our methane car a year or so after mapping Washington, D.C. This time, we're in New York City with Bob Ackley, another Red Sox–loving Bostonian, who works for Gas Safety, Inc. We're parked as close as we can get to Times Square. Our anemometer mast stands tall like the skyscrapers lit through the windshield; our car reflects the cosmic colors of surrounding billboards. A giant green face smirks down to remind us that the problem we're trying to solve is *Wicked*.

We finished mapping Washington, D.C., and found six thousand gas leaks across its streets and twelve instances of explosive concentrations in manholes. The *Washington Post* tracked explosion risks and, based on our data, published an app that allowed readers to see the gas leaks near their homes by entering their address. Capitol Hill—one of the oldest neighborhoods in the city—was flush with leaks. I cannot tell you how many "passing gas" and "hot air" jokes I heard at the expense of the U.S. Congress.

For our next study, we wanted to map cities that had already replaced most of their older cast-iron and unprotected steel piping. If older pipes were the culprit, the density of leaks per mile would drop substantially in cities that had eliminated them. In Cincinnati, for example, the utility delivering gas (Duke Energy), the Public Utility Commission of Ohio, and other partners had steadily replaced most of their older network with new plastic and corrosion-resistant steel pipes. We matched Cincinnati with Durham, North Carolina, another city where most of the older pipes had already been replaced. We then compared leaks across Durham and Cincinnati to cities with much older networks and slower replacement rates, including Boston, Washington, D.C., and Manhattan—hence our visit to Times Square.

Ping! Manhattan looked a lot like Boston and Washington, D.C., with about one leak every quarter mile of road. In Cincinnati and Durham, though, we drove two and five miles, respectively, on average before finding a leak, reductions of 90 and 95 percent compared with Boston, D.C., and Manhattan.

A decade ago, I might have ended the story there: "Scientific data can help us find and reduce emissions of methane and other greenhouse gases." But the story is still unfolding. In the decade since we started our work mapping leaks, the world drove another half a trillion tons of CO_2-equivalent gases into the atmosphere. We are already kissing the bumper of dangerous 1.5°C global temperature increases.

As a result, Phillips and I view climate solutions differently today than we did a decade ago. "I felt compelled to be more active politically and not to check my citizenship at the door just because I'm a scientist," he said. A few years ago, Phillips was arrested protesting construction of the Weymouth Compressor Station that connected two regional gas pipelines. He opposed the station based on its climate harm, its effects on local air pollution and water quality, and because the new station would further harm a community whose residents already bear a legacy of contamination from a petroleum terminal and other industries. Continuing this practice perpetuates the kinds of environmental injustices that Catherine Coleman Flowers highlighted in Chapter 1—and that the state of Massachusetts acknowledged in its evaluation of the project. The proposed site also contained coal ash and other buried toxins whose dust could put residents at risk during construction.

The compressor station was ultimately approved by the state's Department of Environmental Protection after—in Phillips's view—a rushed approval process that included withholding a 759-page document on air pollution data until the last minute. Phillips said, "At that point I felt, 'Wow, the system is just really, really broken.'"

He kept fighting. He launched a hunger strike with three community-based calls to reduce harm: install and maintain air-quality monitors at the site; decontaminate all dump trucks leaving the site; and test for asbestos as the station's foundation was excavated. When I asked him what happened in the end, he said, "The Massachusetts Department of Environmental Protection installed the air-quality monitoring station—addressing one of my demands—but not the other two. The station continues operating today."

Events quickly justified Phillips's concerns. The Weymouth Compressor Station had three "unplanned gas releases" into neighborhood air during its first eight months of operation through 2021.

Our views on gas street leaks have also changed since our initial work. Phillips now promotes "triage and transition" over "replace and rebuild" for pipeline safety. Triaging means repairing the largest and most hazardous leaks as cheaply as possible—using joint-sealing robots, for instance, or lining existing pipes from the inside to seal them. Transitioning means not paying millions of dollars to replace gas pipes that lock in further gas use for decades to come. We shouldn't pay to replace gas pipelines when what we need is a managed, just transition away from gas to cleaner electric.

In the fall of 2023—more than a decade after our initial work—I returned to Boston and met Phillips outside the Charles-MGH trolley "T" stop at the eastern edge of the Longfellow Bridge. He cycled up wearing a brown "Climate Summer" T-shirt with its spoked bicycle-wheel logo.

We hugged and quickly smelled gas. "Look up," he said as we crossed the street. All I saw was the white stucco bottom of the bridge supporting the Red Line metro track. "There," he said, pointing. I finally saw it—the picture-perfect imprint of a manhole cover pressed into the white stucco of the bridge thirty feet overhead.

It took me a second to comprehend. "Gas explosion?" I asked. "Gas explosion," he said, "several years ago." A quick search on social media showed road closures outside of Massachusetts General Hospital and videos of fire shooting out of the manhole like dragon's breath. Leaked gas had puddled below the street, sparked, and exploded, blowing the manhole cover into the bottom of the bridge so hard that it left a perfect imprint.

We inserted our gas probe into the same manhole and immediately measured a gas leak. Rather than fixing the leak—even after an explosion—the utility instead replaced the solid manhole cover with a slotted version that vents leaked gas into the air more quickly. It's a Band-Aid solution that reduces gas buildups and explosion risks but does nothing to stop the leak.

We walked east on Cambridge Avenue, probing for gas leaks as we went. We passed a yellow mark-out scribbled on the street, noting the presence of a twelve-inch cast-iron pipe below. "This cast-iron gas main leaks up and down Cambridge Street," Phillips said. "Gas seeps out in all directions and passes within feet of the manhole that exploded."

In fact, we found some of the same leaks and leak clusters we'd mapped more than a decade before. Stopping when we smelled gas at 100 Cambridge Avenue, we marked a cluster of leaks in the manholes on the south side of the street. We were standing outside the headquarters of the Massachusetts Office of Energy and Environmental Affairs. The office oversees the Department of Public Utilities, the agency that regulates gas pipeline safety across Massachusetts. Progress has been incremental, at best, even literally on the front doorstep of the agencies responsible for the state's energy and public safety.

Recent estimates suggest more than six hundred thousand gas leaks plague U.S. streets alone. They—and the millions of leaks from gas appliances in our homes—illustrate how intractable it is to cut emissions one by one. We need Phillips's "transition"—eliminating gas use and all of the carbon dioxide emissions and other pollution that arise from it. The only way to do that completely is to stop burning fossil fuels, electrify our homes and buildings, and clean up the electrical grid supplying them. Massachusetts eliminated all major coal-fired power around 2017, but natural gas still powers two-thirds of its in-state electricity generation. Despite what the ads say, gas isn't clean, it's just cleaner than coal.

Even our environmental wins are too slow. Washington Gas started its PROJECT*pipes* program in 2014, the year we published our map of six thousand leaks across D.C. The D.C. Public Service Commission approved Phase 1 of PROJECT*pipes* from 2014 to 2019 for $110 million. Seven years later, through Phases 1 *and* 2, the utility had replaced 28.6 miles of cast-iron pipes across its twelve-hundred-mile network. That's a cast-iron replacement rate of four miles per year.

When the world is on fire, I can't celebrate laying four miles of new pipe a year.

What would real success look like?

7

CFC Repair

The methane leaking from our kitchens, cows, and city streets is a triangular pyramid with a carbon at its center and four hydrogens at the points: CH_4. The carbon-hydrogen (C-H) bonds in methane absorb long-wave radiation and bounce like hyperactive schoolkids. These vibrations are how greenhouse gases warm the earth.

A methane molecule can be toyed with, snapping atoms in and out like Lego blocks. Swap methane's four hydrogens for two chlorine and two fluorine atoms ($CH_4 \rightarrow CCl_2F_2$) and you get chlorofluorocarbon (CFC)-12, also known as Freon-12. It was the world's most common refrigerant before the 1990s and the basis for car air conditioners; it also fizzed as the propellant in cans of Silly String.

If, instead, you replace methane's four hydrogens with three chlorine atoms and one fluorine (CCl_3F), you have CFC-11, the next most common refrigerant. CFC-11 was used to produce insulating foams, and it cooled the world's first self-contained air-conditioning unit in a home: Carrier's "Atmospheric Cabinet," introduced in 1932.

Along with being ozone-destroying compounds, CFCs 11 and 12 are super-powered greenhouse gases. They're long-lived and five to ten thousand times more potent than carbon dioxide in the first century after release. Once emitted into the atmosphere, CFCs drift for years—inactive except as greenhouse gases—until lofting into the stratosphere. There, above the ozone shield, they slowly break apart over decades when bombarded by the sun's UV rays, releasing free chlorine and fluorine into the air. CFCs 11 and 12 are the primary causes for global ozone depletion.

When the ozone hole first opened in the early 1980s, British scientists in Antarctica were so skeptical of their measurements that they replaced their instrument to make sure it wasn't broken. Confirming their results with another analyzer, they rocked the world by announcing ozone depletion in 1985, raising the specters of cancer and catastrophe. Stratospheric ozone is Earth's sunscreen, protecting all life from the sun's most harmful rays. Only two years later, dozens of countries ratified the Montreal Protocol on Substances That Deplete the Ozone Layer, the landmark international agreement that limits the production and use of CFCs and other ozone-depleting chemicals.

Ratified today by every country in the United Nations, the Montreal Protocol limits what ozone-depleting gases can be used and sets the schedules for phasing them out. The U.S. Environmental Protection Agency estimates the protocol prevented 440 million cases of skin cancer and two million deaths for Americans born in the twentieth century that would have come from greater exposure to ultraviolet (UV) radiation. It has saved countless more lives globally and many species from extinction.

Just as with the Paris and other climate accords today, U.S. support for the Montreal Protocol in the 1980s hung in the balance. In 1987, then secretary of state George Shultz pushed President Reagan and the rest of his cabinet to support the treaty. Secretary of the Interior Donald Hodel famously suggested that, instead of the U.S. banning ozone-depleting gases, people could wear hats, sunglasses, and sunscreen to counteract the loss of the Earth's ozone shield and protect us from extra UV radiation. Ridicule ensued.

A few years ago (and a year before he passed away to negotiate with the true higher-ups), Secretary Shultz and I met in his office. I wanted to understand why countries acted so quickly on the ozone hole and what lessons might apply to climate action. A spry centenarian, he turned in his leather-backed chair as I entered. An oversized globe stood to his left. A framed medal hung on the wall above it—Honorary Officer in the Order of Australia—just one of the many awards he could have changed on his wall daily, like his clothes.

On this mundane Thursday Shultz was better dressed than I was on all but my wedding day. He sported a mustard jacket, deep blue vest, striped collared shirt, and judicious ornaments: wedding band, maroon H pin for the Hoover Institution, and matching handkerchief sprouting from his jacket pocket like a rose petal. Bygone admonitions from my mother to sit up straight flushed like doves from a magician's sleeve.

Both dove and hawk, Shultz was a world leader in reducing the dangers of nuclear weapons and in pricing carbon pollution. He helped broker the INF, the Intermediate-Range Nuclear Forces Treaty with Russia that limited ballistic and cruise missiles and made the world safer for decades (until the United States withdrew from the treaty in 2019). He also coauthored an influential plan for carbon pricing—"The Conservative Case for Carbon Dividends"—to reduce carbon dioxide emissions in the fight against climate change.

Now seated ramrod straight, I asked Secretary Shultz why the Montreal Protocol succeeded in the United States and elsewhere. He began with the context for the treaty, crediting his assistant secretary, John Negroponte, for convincing him and other American leaders of the situation's gravity.

"The Montreal Protocol stands out as maybe the only effective environmental treaty that the world has accepted," Shultz said. "A small group at the State Department and EPA became convinced the ozone hole was real and was a big problem. John kept briefing me on it and I became convinced he was right.

"So I had two private meetings with President Reagan in a week, and I brought it up with him. We talked about it in some detail." He described Reagan as an "instinctive environmentalist" who enjoyed working outdoors at his ranch near Santa Barbara, California. "This got to him. The more we talked, the more he became convinced it was a big problem. Then he did the opposite of what people do today. He went to skeptics and said, 'You don't agree with us and I respect that. But you do agree that if it happens, it's a catastrophe. So why don't we take out an insurance policy.'" Shultz highlighted the need for "insurance" and caution often in our discussion—not just for ozone depletion, but for climate change and weapons proliferation, too. He said, "The insurance policy concept appealed to people. It didn't get them on our side, but it got them off our backs."

I asked Shultz about barriers to action. He credited "R&D" (research and development by companies) for quickly finding replacements for CFCs as refrigerants and solvents. He also credited Reagan again and bipartisan support in the U.S. Senate, which was controlled at the time by a Democratic majority: "When the President is clear and convinced of something, a lot of opposition goes away." How quaint this view seems today. The treaty was ratified unanimously by Senate vote in March of 1988.

The ban on CFCs 11 and 12 mandated by the Montreal Protocol gave countries different deadlines to act. The ban for wealthier countries began in 1996, allowing almost a decade for an orderly transition. Poorer countries got even longer, until 2010, to replace their coolants and infrastructure. So did manufacturers of healthcare products such as the asthma inhalers where a safe alternative wasn't yet available. Practically speaking, though, U.S. production had already peaked in the 1970s over concern for the ozone layer, even before the discovery of the hole in the ozone shield shocked the world. By 2010, all global production of CFC-11 and 12 ended.

Well before the date of the final ban in 2010, then, proactive reductions drove atmospheric concentrations of both gases to peak and begin to inch downward. The key word is "inch." Both gases were, and are, heading to zero slowly. Though neither gas forms naturally, both are long-lived, remaining in the air for 45 and 110 years on average for CFCs

11 and 12, respectively. Consequently, recovery or atmospheric restoration will occur decades to centuries from now even after—and if—a successful ban on CFCs is maintained.

If you know the lifetime of a gas and the processes that destroy it, and you know the rate at which it enters the atmosphere—close to zero for CFCs, with minor emissions from infrastructure still in use—you can predict its decline through time and when it will finally be gone.

For a decade or two things went swimmingly. Concentrations of CFC-12 continued moving downward year by year at the pace predicted by natural destruction from sunlight in the stratosphere. But around 2013, CFC-11 concentrations diverged from their expected path, no longer dropping as fast as expected.

Scientists were confused. Had something changed in the stratosphere, perhaps the appearance of a different chemical that altered the life span of CFC-11? If that were the case, concentrations near the Earth's surface would be similar everywhere because no one was supposed to be releasing the gas and because it's mixed evenly throughout the atmosphere. In contrast, if someone had started illegally making or using CFC-11 again, you would discover hotspots near the surface, ripples like pebbles dropped in a pond. If you had good monitoring tools and were smart enough, you could tell whether someone was cheating—and where.

Enter detective-scientist Matt Rigby. An auburn-haired chemist at the University of Bristol, with a matching close-cropped beard, Rigby sports a lab coat instead of a gumshoe's trench coat and reads spectroscopic outputs instead of crime scenes. Rigby and I do methane-related research together through the Global Carbon Project.

Rigby felt a sense of mission about his work on CFCs. He said, "I'd done a PhD in atmospheric physics, and I was keen to apply what I'd learned to a global problem with relevance to society."

Rigby, along with Steve Montzka at the U.S. National Oceanic and

Atmospheric Administration (NOAA) in Boulder, Colorado, and other colleagues, were the first to recognize that CFC-11 concentrations weren't dropping as fast as they should have been. Using their global monitoring network, they found higher concentrations of CFC-11 in the northern hemisphere starting in 2013, something not seen for decades since CFC production was supposed to have stopped. Just as importantly, the declines in other long-lived gases, such as CFC-12, *hadn't* changed, which ruled out some fundamental difference in atmospheric chemistry or circulation that would explain the strange CFC-11 results. Their research suggested an illegal source of CFC-11 pollution north of the equator.

Using data from the Mauna Loa station in Hawaii, Rigby and his colleagues found higher CFC concentrations and other industrial pollutants when the air blew from the west. Through careful analyses they concluded that the increased CFC-11 emissions likely came from eastern Asia. They also proposed that the unusual CFC-11 measurements arose from new production that violated the Montreal Protocol—someone was cheating.

With evidence of a crime in hand, Rigby searched for suspects. Once you narrow a source to a particular region, different tools can help you find it. You might put a sensor on an airplane and look for hotspots; when you find one, you keep flying upwind until you locate it, circling lower and lower to isolate the source. Just upwind the signal should disappear; just downwind, you'll detect it again. You can even measure how much is being released by tracking how big the plume of air is, how quickly it's moving, and how large the increase in concentration is upwind and downwind of the source. We apply this approach today in our work measuring methane gas leaks in oil fields using planes and helicopters.

Rigby needed a different approach, though. He lacked permission to fly in China to sleuth out CFC-11 polluters. In a follow-up study with a larger team of scientists, including regional modeler Anita Ganesan (Rigby's partner at work and in toddler parenting), they analyzed two regional stations about five hundred miles apart north to south and much closer to China than Mauna Loa. One was the Gosan Station on Jeju Island, South Korea, and the other was on Hateruma Island, Japan, stations run by Rigby's colleagues Sunyoung Park and Takuya Saito.

Imagine an instrument perched on a tower with a flagpole. As the wind direction changes, a pollution source will only be detected when the wind blows from the upwind source to the flagpole. The rest of the time the observations will be normal. High gas concentrations wink in and out as the wind changes direction and blows polluted air briefly across the sensor.

Through time, you can narrow the direction of the source, drawing a line or wedge on a map upwind from the flagpole. But now the second station enters into play. It, too, sees the pollution, but only when the wind blows from a different direction. You can draw a second line upwind from that sensor, too. Just as in geometry class, you find the approximate location of the source where the two flagpole lines intersect (though Rigby uses atmospheric models rather than compasses and protractors).

The lines for CFC-11 intersected at Shandong Province, China, and multiple lines of evidence confirmed the region as the source. If someone was manufacturing CFC-11 illegally in eastern China, for instance, you would also expect a parallel jump in emissions of the ingredients used to make CFC-11. Taking measurements of carbon tetrachloride—the main precursor for producing CFC-11—at the same monitoring stations in Korea and Japan, Rigby and his colleagues found that X marked the same spot. They measured higher concentrations of the raw materials there, too, describing "a new source or sources of emissions from China's Shandong province after 2012."

"When we saw very similar numbers for the rise in emissions from the Korean station alone, that was a lightbulb moment," he said. "We could explain most of the global rise in CFC-11 from just one place in China."

Despite his shocking discovery, Rigby believed the long-term damage of this cheating would be modest if China eliminated the new emission sources quickly. If not, then the cheating would "undo part of the success of the Montreal Protocol." The race was on to find and eliminate China's illegally produced CFC-11.

If you're thinking to yourself, "Well, of course: China," not so fast. It's true that many environmental rules there lack the punch of those in the United States and Europe. However, China has reduced its air pollution substantially over the past decade, investing in renewables and natural gas to replace dirtier coal-fired electricity. China is also deeply concerned with international perceptions of global stewardship and leadership. It knew it had a problem.

External investigations and Chinese documents both highlighted a major source of CFC-11: the manufacturing of polyurethane foam used to make insulation. Before the ban, manufacturers used CFC-11 as their blowing agent for making foam insulation. There's irony here, because the foam made in China was being used to upgrade building insulation and to make refrigerators and other appliances bought in Europe and the United States more energy efficient. Reports by the nonprofit Environmental Investigation Agency documented CFC-11 use by eighteen companies across multiple Chinese provinces, with companies admitting using CFC-11 instead of replacement chemicals. Laboratory tests confirmed residual CFC-11 in foams from at least three companies after the ban from the Montreal Protocol.

The *New York Times* published an exposé of the industry in November 2018, quoting environmental officials who acknowledged the problem. One Chinese official in Shandong admitted, "Currently there is still a large volume of illegally produced CFC-11 being used in the foam industry."

Chinese business executives, too, were surprisingly frank about the challenges they faced in a competitive industry. Zhang Wenbo, owner of a refrigerator factory in Xingfu, said, "You had a choice: Choose the cheaper foam agent that's not so good for the environment, or the expensive one that's better for the environment. Of course, we chose the cheaper foam agent. That's how we survived." According to the *New York Times* report, Chinese officials entered Zhang's factory soon afterward and ordered it closed. One challenge is that thousands of such small factories may exist in China, winking in and out of business like fireflies.

China has pledged to crack down on the manufacturers of ozone-depleting chemicals. China's minister of ecology and environment, Li Ganjie, said that producers of carbon tetrachloride, the CFC-11 precursor, were now under strict twenty-four-hour supervision. The attention, supervision, and even arrests in China appear to be working.

Newer data suggest the cheating has stopped. In consequence the global decline in CFC-11 concentrations has returned to normal. Matt Rigby, Steve Montzka, and others documented "an accelerated decline in the global mean CFC-11 concentration during 2019 and 2020." They concluded, "If the sharp decline in unexpected global emissions and unreported production is sustained, any associated future ozone depletion is likely to be limited."

I asked Rigby whether he'd seen evidence of reduced CFC-11 production in China and what he thought would happen there. He expressed excitement and—like George Shultz—the need for vigilance. "We need to be cautious, but the signal seems to be going back to where it was in 2012, like something has switched off. If that turns out to be true, it's such an exciting thing for us." The pace of recovery was surprising, too. He added: "What really struck me was how rapidly things turned around. Over the course of a year, essentially, the emissions dropped by about the same amount they'd grown in 2013."

Rigby continued with what scientists dream of saying: "We're all small parts in this, but it's gratifying that something we've done made such a difference." He added, "It was fantastic to see concentrations start heading down again—it was amazing. Coincidentally, we also saw the decline in pollution at the Gosan station in Korea, which confirmed an emissions decline from China."

Rigby did express regret about one aspect of his work: the years of delay between the initial uptick in CFC-11 emissions and when they published their first paper. He mused, "Oh, my God, how on earth did I miss that? We had those signals in our data since 2013." He credits his colleague Steve Montzka for first noticing something was amiss. It took years, though, to confirm the signal and to wrestle with the potential political implications of being right (or wrong). Once they did notice, they had to be certain the signal was real and that it persisted, balancing caution with the need to provide timely information. In retrospect—and given the enormous consequences—Rigby was right to be cautious. Nevertheless, he wished he'd moved faster: "Science moves slowly sometimes."

The atmosphere moves slowly, too, especially for long-lived gases such as CFC-11 and -12. The atmosphere will be free of both gases only in a century or two, and only if emissions remain near zero. Restoring the atmosphere takes a long time for some gases; either we wait for them to be cleansed from the air naturally or we pay to remove them actively. For gases such as CFCs, found at only parts per trillion, paying to remove them is impossible. So we wait, heeding Matt's call for vigilance: "There's a lot of work still to be done to ensure that the Montreal Protocol keeps succeeding as an environmental treaty. And solving the climate problem is even more complex than solving ozone depletion."

But even for long-lived gases, our actions have borne fruit today. By October 2022, Antarctica's ozone hole was smaller than it had been in decades. It had shrunk by more than a third compared to its peak ten or twenty years ago. A NASA statement concluded "The elimination of ozone-depleting substances through the Montreal Protocol is shrinking the [ozone] hole."

An extra benefit of the Montreal Protocol is how much warming the world avoided by phasing out CFCs, acknowledging that their current replacements are also powerful greenhouse gases. According to one recent study, some Arctic regions would have been as much as 2°F warmer today without the ban. The benefits for the future are eye-popping—almost 2°F of average global warming will have been avoided by 2050 and the Arctic will have dodged an additional 6°F of warming.

The ozone hole will still appear every fall for decades, probably until at least 2075. It's shrinking, though. George Shultz's view of the Montreal Protocol as a "prudent approach to risk management" was spot-on.

Not all halocarbons live in the air as long as CFCs 11 and 12. Methyl chloroform's lifetime in the atmosphere is only about six years—roughly half that of methane's—so emission cuts quickly lead to decreased concentrations. Until the Montreal Protocol banned its production in 1996, methyl chloroform was the Swiss Army Knife of industrial chemicals: solvent, cleanser of circuit boards and households, metal degreaser, adhesive, insect fumigant, and cleaner of choice for photographic film and negatives. Its global concentration peaked at around 150 parts per trillion before falling to today's value of only 1 part per trillion.

Unsurprisingly, given his expertise, Matt Rigby and colleagues also track methyl chloroform in the atmosphere. Rigby told me: "If you wanted to show one figure that most vividly demonstrates how successful the Montreal Protocol has been, I think methyl chloroform is hard to beat. And because its lifetime is short compared to the other ozone-depleting substances, we've already seen this dramatic decline in the atmosphere."

One ozone-depleting gas, then, already illustrates what atmospheric restoration will look like for methane—first—and then other greenhouse gases later. To make this come true, we'll need to zero out emissions from every sector of the economy.

If we don't, we enter the netherworld of Drawdown.

Part II
DRAWDOWN LAST

8

Drawdown

C umulative emissions matter the most for longer-lived gases such as carbon dioxide and nitrous oxide. Cumulative emissions of carbon dioxide since 1850 are about two and a half trillion tons— from all the coal mines ever dug, all the oil and gas ever burned, and all the net losses of carbon in trees and soils from deforestation, agriculture, and other changes on land. A trillion tons is an unfathomable number—about three million Empire State buildings worth of mass.

Restoring the atmosphere in a lifetime—returning concentrations to preindustrial levels—is impossible for carbon dioxide. It would mean removing a trillion tons of pollution now in the air. No one reading this book will live long enough to see that happen; the quantities are too vast, today's emission rates are too high, and carbon dioxide lasts too long in the atmosphere. The same is true for nitrous oxide.

But for methane we need to remove "only" two or three billion tons to restore the atmosphere to preindustrial levels. (In comparison, three billion tons is roughly one month's worth of current carbon dioxide emissions.) My dream is to see this happen in my lifetime.

As discussed earlier in this book, if we want to slow global warming by reducing greenhouse gas concentrations over the next decade or two, reducing atmospheric methane concentrations is the best—and perhaps only—lever at our disposal to shave peak temperatures and delay crossing critical temperature thresholds, such as 1.5 and 2°C global increases.

Most methane is cleansed from the atmosphere naturally within a decade of its release, split by sunlight and nature's detergents. Because of this short lifetime, if we could eliminate all human-caused methane emissions from agriculture, fossil fuels, and fire—a huge if—methane's concentration would return to safer preindustrial levels within a decade or two. If that doesn't happen, we will need to actively remove it from the atmosphere to reduce peak temperatures.

In the 2010s, methane was responsible for about 0.5°C of global surface warming compared to the late 1800s. Carbon dioxide contributed a little more, about 0.75°C warming over the same period. Other gases warmed the Earth, too, particularly nitrous oxide (almost 0.1°C), with another 0.1°C warming coming from refrigerants such as the chlorofluorocarbons (CFCs), which cause the ozone hole, and their replacements, hydrofluorocarbons (HFCs).

Carbon dioxide is two hundred times more abundant in the atmosphere (425 ppm) than methane (~ 2 ppm). However, pound for pound, methane is eighty or ninety times more potent two decades after its release and roughly thirty times more potent than CO_2 after a century. Concentrations of methane have also increased far more in the Earth's atmosphere since preindustrial times (160 percent) than those of carbon dioxide ("only" 50 percent).

Although carbon dioxide remains the most important greenhouse gas, methane is the second most important and the priority of my current climate work. No other greenhouse gas provides the opportunity for such rapid atmospheric restoration. Restoring the atmosphere will require drastic cuts in greenhouse gas emissions and—most likely—the removal of greenhouse gases already in the air.

The need for "drawdown," defined here as removing carbon dioxide, methane, and other greenhouse gases from the atmosphere *after* their release, arises from failure. We have flooded the atmosphere with trillions of tons of carbon dioxide pollution—most of it in the last fifty years—even after the danger to life was clear. In fact, annual global fossil carbon dioxide emissions rose 60 percent since publication of the first IPCC report in 1990 that documented the climate problem comprehensively. We have not just failed but failed spectacularly.

Given our failure to act, we've left future generations little choice but to clean up after us if global temperature increases are to stay below 1.5°C or 2°C thresholds. And paying to remove greenhouse gases from the air tomorrow will cost much more than preventing them from entering the air today.

Can drawdown technologies actually work? They can. Can they work at the billion-ton scales needed? Perhaps, but only if someone pays enough for them. They aren't magic—as we'll see—and they're expensive.

Under almost every scenario, meeting the 1.5°C temperature target will require removing some previously emitted carbon dioxide (CO_2) from the atmosphere. One of our recent Global Carbon Project analyses led by Sabine Fuss in Germany concluded that if we could keep cumulative global emissions below 750 billion metric tons (about two decades of current emission rates) before 2100, about 400 billion tons of carbon dioxide would still need to be removed from the atmosphere to keep global temperatures increases below 1.5°C in 2100.

Even if we estimate the cost of removing carbon dioxide in the future as $100 per ton, an aspirational target at best and five times lower than a typical industrial cost today, removing 400 billion tons will cost $40 trillion—larger than the combined annual GDPs of China and the United States. Younger generations are legitimately asking: "Why should *we* pay for this?"

Even if we find a way to pay this $40 trillion bill, carbon removal is tricky to achieve at scale. The atmosphere contains about 1 molecule of carbon dioxide for every 2,500 molecules of other gases (0.04% CO_2), which makes finding and "removing" carbon dioxide like pulling needles repeatedly from a haystack. In contrast, about one in ten molecules emitted from the smokestack of a fossil power plant is carbon dioxide (10 percent CO_2). Thus it makes no sense to pay to remove dilute CO_2 from the atmosphere while allowing smokestacks to keep belching concentrated carbon dioxide into the air today. Wherever we keep burning fossil fuels, we need to capture the carbon pollution from smokestacks now, before it enters the air. And unless fossil power plants are serving a region of the world mired in energy poverty, no new fossil plants should be built without carbon capture and storage technology.

In 2023 there were only about forty carbon capture and storage (CCS) plants running worldwide, with another 325 in development. The amounts of carbon stored annually rose to twenty-nine million tons of carbon dioxide, less than one-thousandth of all fossil carbon dioxide emissions. Compared to the forty active CCS facilities, thousands of fossil plants operate globally. If all of those fossil plants complete the end of their lifetimes without carbon capture and storage, their "committed emissions" will entail hundreds of billions of additional tons of carbon dioxide pollution, more than enough to push us past 1.5°C and possibly 2°C.

If we fail to curb emissions and fail to capture and store the carbon pollution from large sources, then carbon drawdown or removal technologies come into play. Using land is one of the cheapest options—especially regrowing the carbon in forests and soils lost to the atmosphere through deforestation and agricultural activities.

The world lost a billion hectares of forest in the twentieth century; most of that land is now used for growing crops and cattle ranching. Agricultural activities such as plowing have released billions of tons of carbon dioxide into the atmosphere from the world's soils. These carbon losses from soils and forests underpin natural climate solutions, approaches that stem carbon losses and put carbon back through conservation, restoration, and improved land management. Fairly optimistic estimates suggest such practices could provide one-third of the climate mitigation needed by 2030 to stabilize global warming below

2°C. Natural climate solutions are also currently the cheapest way to off-set fossil fuel pollution, often at costs of approximately $10 per ton of CO_2 stored—acknowledging that carbon stored in a tree is not perma-nent like carbon locked away underground.

We can remove billions of tons of carbon from the atmosphere through natural climate solutions such as forest and wetland resto-ration, tree planting, no-till farming, and other actions. A more plant-based diet, especially eating less red meat (see Chapter 3), would also reduce deforestation, the global tally of a billion-plus cows, and methane emissions, while sparing land for other ecosystems and human uses.

Can we rely mostly on natural climate solutions? No, at least not to offset anywhere near the almost forty billion metric tons of annual fossil carbon pollution, and not if we don't want it to compete with agriculture for land.

Without immediate, drastic emissions cuts, industrial greenhouse gas removal will be needed to keep global temperature increases below both 1.5°C and 2°C. Scientists have studied atmospheric carbon dioxide removal for more than a decade: the twin steps of capturing carbon diox-ide from the air and storing it out of harm's way. Plants, including trees, grasses, kelp, and phytoplankton, as well as some microbes, take up car-bon dioxide through photosynthesis. Rocks and industrial chemicals can also be used to remove CO_2 from air.

Beyond the natural climate solutions discussed above, there is a hy-brid industrial approach known as BECCS, which stands for bioenergy with carbon capture and storage. For BECCS, users gather or harvest trees or other plants that extract CO_2 from the atmosphere, burn the biomass to produce electricity (or convert it into biofuels), and pump the carbon dioxide pollution underground to keep it from returning to the air. Of all the drawdown or negative emissions technologies, BECCS is the only one that provides energy rather than requires it (and, done carefully, can provide close to carbon-free or even carbon-negative en-ergy). A recent U.S. National Academy of Sciences study put the U.S.

potential for BECCS at three to five billion metric tons of carbon dioxide removed per year without large adverse impacts. That amount is roughly one-third to one-half of the carbon removal many scientists believe will be needed each year globally to maintain a livable climate. We will explore BECCS in the next chapter.

Another drawdown technology is enhanced weathering. This approach tries to accelerate the rate at which rocks react naturally with atmospheric CO_2. Igneous basalt—dark-grained and volcanic—covers a tenth of the Earth's continental surfaces and most of the ocean floor. Basalt contains lots of minerals that react with carbon dioxide to form carbon-rich rocks. Calcium carbonate, or common "limestone," for instance, combines one calcium atom with carbon dioxide and an extra oxygen ($CaCO_3$). It is the stone that was used to build the Empire State Building and the Great Pyramid of Giza.

Imagine mining basalts, crushing them, and exposing them to the air to react with carbon dioxide. You might even fertilize an agricultural field with them, enhancing plant growth because of the extra calcium, magnesium, and additional nutrients the rocks release. Alternatively, you might just expose crushed rock to air and rebury it after it reacts fully with atmospheric CO_2. Cost estimates for enhanced weathering are $75–250 per ton of CO_2 removed. Start-up companies are forming to try, but enhanced weathering hasn't scaled to commercial-level projects yet. We know weathering works in nature over thousands of years; can companies speed up the process enough for it to work in only years?

Finally, dozens of companies are working on direct-air capture (DAC) of carbon dioxide using specialized chemicals, as we will see in a journey to Iceland in Chapter 10 to examine a process that combines aspects of enhanced weathering and DAC. Nitrogen-based "amines" have been used for decades in refineries and chemical plants to scrub carbon dioxide from gas streams. Hydroxides are a second family of chemicals used in commercial direct-air capture operations. In both cases, the original chemicals can be regenerated using heat or by changing the acidity of a solution. Concentrated carbon dioxide is released during this chemical regeneration.

In most direct-air capture operations, the captured carbon dioxide must be pressurized and pumped underground, just as for BECCS during the carbon capture and storage (CCS) part of the process. The current cost range for direct-air capture is about \$250–600 per ton of CO_2 removed, far more than for natural climate solutions. Today, companies are removing a few tens of million tons of carbon dioxide a year from the atmosphere industrially. That's a start but far from the billions of tons needed yearly.

Beyond carbon dioxide, we'll need to remove other greenhouse gases from the air, too. Two-thirds of global methane emissions come from human actions, including fossil fuel use and agriculture. Global methane concentrations are now 2.6 times higher than they were two centuries ago.

Removing methane from the atmosphere is more difficult than removing carbon dioxide. It's less abundant in air than carbon dioxide and therefore more difficult to isolate. But methane removal also has advantages. Notably, you don't need to capture it and pump it underground. If you can react it using catalysts or nature's detergents, you can convert it to CO_2 and release it, eliminating the capture and storage phases that make carbon capture so expensive. Because all methane emitted into the atmosphere eventually becomes CO_2, methane removal simply speeds nature's reaction. Trading methane for carbon dioxide is good for the climate because methane is so much more potent than CO_2.

If feasible at scale, methane removal could help shave tenths of degrees off peak temperatures and buy us more time to cut carbon dioxide emissions further before a given temperature threshold is passed. Further, some scientists believe that it is possible—even likely—that catastrophic levels of methane could be released from Arctic permafrost and tropical wetlands this century because of warming temperatures. Preparing for methane removal today could provide much needed insurance against future disaster.

For all drawdown solutions, we will need either a regulatory mandate or a global price or market for carbon and methane pollution to prompt action. An "upstream" carbon price adds a fee wherever fossil fuels are extracted, with the extra cost passed on to consumers in the price they pay for products derived from fossil energy. (Discussions are needed on what to do with the proceeds from fees and how to keep poorer people from paying higher prices for their energy.) This price would better shift the financial burden of emissions onto the people, companies, and industries responsible for them and would more closely reflect the real cost of fossil pollution. None of the options I've discussed above is feasible at large scale without a carbon market or, short of that, policy mandates requiring action.

Unfortunately, a national carbon price is unlikely anytime soon in the United States. The last market-based climate bill with any chance of passage was more than a decade ago: the Waxman-Markey American Clean Energy and Security Act of 2009, which passed the U.S. House with bipartisan support but never reached a vote in the Senate. Getting a market-based climate bill through either the U.S. House or Senate these days seems a tall order, despite the success of such bills to reduce acid rain cheaply and efficiently in the past.

Although the cost of drawdown is high, the cost of doing nothing is staggering. Insurance companies understand costs and risks better than anyone. Insurance giant Swiss Re, the world's second-largest reinsurance company (reinsurance companies insure insurance companies), recently estimated that the global economy could shrink by 18 percent if no climate mitigation action is taken, at a cost of up to $23 trillion annually by 2050. Their report concluded: "Our analysis shows the benefit of investing in a net-zero economy. For example, adding just 10% to the U.S. $6.3 trillion of annual global infrastructure investments would limit the average temperature increase to below 2°C. This is just a fraction of the loss in global GDP that we face if we don't take appropriate action." The insurance industry believes that the costs of climate inaction dwarf the costs of climate action.

To reduce these costs, we need to cut emissions—and then cut them some more. (That's the reason I began this book with "Emissions First.") We'll also need to deploy energy technologies that not everyone reading this book will like. If we shun nuclear power, we might need to harvest trees and biomass for energy from additional land the size of Texas, changing land-use habitats and conservation priorities. If we scuttle those options, we might need thousands of natural gas plants pumping pollution underground, monitoring groundwater beneath them to make sure the pollution stays put. And if we don't do that, we might pay far more and cover a tenth of usable land area with solar panels to keep the lights and heat on during winter, as our recent analysis showed for the state of California. We need a new energy system that is reliable and capable of reducing emissions, and, as in the classroom and workplace, diversity helps.

We will need diverse drawdown technologies, too. We'll need to implement natural climate solutions, restoring forests and soils wherever possible. We need to lower the cost of drawdown technologies and hope people accept such technologies at thousands of locations. We need to discuss personal issues of population, diet, energy use, and inequality.

In truth, I'm frustrated writing about "drawdown" technologies because we shouldn't need them. I've watched years of climate inaction roll by like floats in a parade. When will the victory parade finally begin?

9

Out of Gas

New fossil smokestacks still sprout like corn plants in May. Though withering in the United States and Europe, coal plants keep budding globally. The world added hundreds of new coal-fired power plants in 2022, mostly in China, India, and Asia more broadly. In contrast only 110 coal plants worldwide were retired in 2022—for a net *gain* of twenty-five billion watts of coal-fired capacity globally.

Utilities in the United States have not built a new coal-fired power plant since 2015, but the U.S. still has the third most coal capacity of any nation. There is probably no greater climate priority today than using renewables to replace the thousands of coal plants still operating globally, including more than four hundred U.S. plants alone.

In fact, no fossil fuel shows a sustained decline in global use—not coal, not oil, and certainly not methane-rich natural gas. Each fossil fuel is responsible for billions of tons of carbon dioxide pollution a year when burned, and each fuel leaks methane during extraction and use.

And at this point, to maintain a safe climate, every ton of carbon pollution released into the air will need to be removed later—paid for by someone, somewhere. Building and maintaining fossil power plants without capturing their carbon dioxide at the point of pollution makes no sense because keeping greenhouse gases out of the air today will *always* be cheaper than removing them after they've entered the atmosphere and diluted throughout it.

Burning a fossil fuel for energy and capturing and storing the carbon pollution would be a major improvement over what most countries do today—which is to burn fossil fuels and pollute for free—but we need to do even more than that. We need to remove greenhouse gases from the air, too. The best we can do by combining fossil fuels and carbon capture and storage technologies is to get close to, without ever quite reaching, zero-carbon energy. Other potentially carbon-neutral approaches include growing plants for biofuels, such as ethanol, that replace fossil gasoline. In principle you don't have to capture the carbon dioxide after burning the biofuel because the carbon in the biofuel starts in the air, yields energy when the biomass is burned for electricity or transportation fuels, and then returns to the air where it began.

However, to *remove* carbon dioxide from the air we need to combine both approaches. We can't start with a fossil fuel as the energy source because we'd be releasing geologic carbon into the air. If we could grow crops or trees to remove carbon dioxide from the atmosphere through photosynthesis, burn them to generate energy, and then store some or all of the carbon dioxide pollution underground, that would both remove carbon dioxide from the air and yield energy. Ideally the plant biomass burned to produce the energy would come first from leftover waste biomass rather than from healthy forests that we cut. As described briefly in the previous chapter, this process is called bioenergy with carbon capture and storage (BECCS).

Through BECCS, carbon that starts in the air ends up belowground, reducing atmospheric CO_2 concentrations. Of all the drawdown or negative emissions technologies, BECCS is the only one that provides energy instead of requiring it.

As with any technology that pumps pollution underground, BECCS requires monitoring the reservoir for decades to make sure the carbon dioxide stays put. Still, BECCS is relatively cheap by negative emission standards and BECCS plants are already running commercially today.

Biofuels for transportation is another form of BECCS. Although you can't capture the carbon dioxide pollution from a car's tailpipe today, you can capture the CO_2 produced while manufacturing biofuels such as ethanol, which moves some carbon captured from the atmosphere underground. Biofuel facilities around the world release pure carbon dioxide into the air while producing ethanol and other transportation fuels. U.S. manufacturers make more than half of the world's ethanol fuels, largely because the U.S. government mandates that ethanol be mixed into gasoline. U.S. ethanol production releases 40 million tons of carbon dioxide into the air yearly as the biomass ferments. Why not capture that CO_2 before it enters the air?

Making ethanol for cars and trucks is like brewing beer. You start with sugar, typically barley starch for beer or—to make ethanol—corn in the United States or sugarcane in Brazil. You ferment the mashed-up plants—using yeast to turn the sugars into ethanol—releasing carbon dioxide as the microbes break down larger carbon molecules. If you've ever home-brewed beer, you've watched with satisfaction as carbon dioxide bubbles out of your carboy, confirming that your yeast colony is active. The yeast in industrial ethanol vats also produces pure CO_2, so it makes sense to capture the millions of tons of CO_2 released through biofuel fermentation before it bubbles away. But someone has to pay to capture carbon pollution and pump it deep underground.

Engineers at agri-giant Archer Daniels Midland (ADM) in Decatur, Illinois, are currently running the largest BECCS plant in the world. How does it work? And how did they pull it off against long odds?

ADM's carbon storage facility sits next to their 350-million-gallon-a-year ethanol production facility, surrounded by hundreds of miles of cornfields. To understand the company's goals and approaches for the project, I spoke first with Alison Taylor, ADM's chief sustainability officer, and then to ADM engineer Scott McDonald, who runs biofuels production and carbon storage at the Decatur site.

Taylor leans in when broaching the changes she's seen in her lifetime and the business reasons for taking climate action. "I grew up in the southern United States watching and sometimes getting evacuated for hurricanes. I was also a marine biologist on the coast of North Carolina. I've seen the way hurricanes have developed—becoming more frequent and bigger—and literally changing coastal features permanently. And I have to say, when I first started thinking about climate change, I thought, 'This is very important for future generations, but I might not see many consequences in my lifetime.' But I was wrong."

Taylor combines training in sustainability and law with federal policy experience. She served as chief counsel to the U.S. Senate Environment and Public Works Committee and was vice president of sustainability for a decade at Siemens, the largest industrial manufacturing company in Europe. She oversees global sustainability strategy at ADM, tracking its progress toward U.N. Sustainable Development Goals 2 and 13: "Zero Hunger" and "Climate Action."

Taylor first gained experience with carbon capture and storage (CCS) technologies while at Siemens. Carbon capture uses chemicals that react with carbon dioxide in air. As mentioned earlier, a step-change in temperature or acidity frees the CO_2 from the chemicals and allows engineers to pressurize and pump the captured carbon dioxide underground. Taylor said, "Siemens had a group dedicated to CCS. It was an important bet for the company, with people hoping there would be a technology to limit pollution from fossil-burning plants."

But inconsistent policy support slowed the deployment of CCS by at least a decade. For example, the FutureGen project, a government-industry partnership, hoped to capture and store a million tons of carbon dioxide emissions a year for two decades from a coal-fired power plant in Illinois. The plant was to be retrofitted to burn its coal in pure oxygen and carbon dioxide (a process called "oxy-combustion") instead of in nitrogen-rich air. Doing so would make it produce a waste stream of almost pure CO_2 suitable for recovery and storage. In 2009, the U.S. Department of Energy committed a billion dollars to the project in stimulus funding (the American Recovery and Reinvestment Act of 2009), with the remaining quarter of costs to come from private companies. However, cost overruns and a lack of private funding commitments led the government to cancel the project six years later.

"When FutureGen ended," Taylor explained, "I saw interest at Siemens fall apart, too. So it's interesting for me now to see this resurgence in CCS. The government is talking about funding these projects again." In fact the U.S. government is doing more than "talking" about funding such projects; the U.S. recently announced $2.5 billion for carbon-management projects such as what FutureGen would have accomplished.

Taylor supports such projects but acknowledges some resistance to them: "Many NGOs are in favor of this technology, but some aren't because they view it as prolonging the life of fossil fuels. I believe it's necessary, that we have to do something quickly. We can't wait for perfect technologies to replace fossil-fuel technologies."

Lead engineer Scott McDonald concurred and jumped in to discuss the history of their BECCS project at Decatur and how ADM made it work. "You have to go back to 2005, when the [Department of Energy] established regional partnerships for capturing and permanently storing greenhouse gases," he said. "The Midwest regional consortium did a lot of work looking at ways to store geologic CO_2 and proposed the Illinois basin as a storage site. In Champaign, they already had gas storage wells drilled into the Mount Simon formation," a massive sandstone reservoir that underlies much of Illinois and Indiana. The Mount Simon is one of the largest and most promising formations for permanently storing carbon dioxide a mile or more underground.

"We began injecting carbon dioxide from our ethanol unit in 2011 and ran the project for three years," McDonald said. "We closed it after we'd successfully reached our target of injecting a million tons." Decatur was the world's first bioenergy carbon capture and storage project.

There's a lot to unpack behind McDonald's comments: How do you pump millions of tons of carbon dioxide underground? Is it safe? And will it stay there?

The "how" is easiest to answer. Companies have been pumping waste underground for decades, particularly when disposing wastewater from oil and gas operations. (Oil wells produce far more wastewater than oil worldwide, including natural radioactive materials in waters that must be disposed of.) About a trillion gallons of brines and other waste fluids are pumped underground each year in the United States in hundreds of thousands of EPA "Class II" injection wells, a toxic legacy of fossil fuel production whose impact few people grasp fully. In the relatively rare cases of well failure, injected wastewater has gurgled up in Los Angeles dog parks, Texas farms, Ohio fields, and Louisiana roadside ditches. Referring to the millions of waste-disposal wells dotting the U.S. landscape, one EPA geologist said, "The United States looks like a pin cushion."

Such wells are also used to pump carbon dioxide underground to recover oil. In some older oil fields, so much oil and water have been removed over many years that the flow of oil slows to a trickle as the pressure underground drops. The goal of enhanced oil recovery (EOR) is to repressurize the basin using CO_2 to push out more oil.

For purposes of reducing emissions, though, only a small portion of the carbon dioxide used in EOR typically stays underground, and most of that carbon is offset both by the extra carbon emitted when the recovered oil is burned and by the extra energy used to capture, compress, and pump the carbon dioxide through pipelines. EOR and carbon storage make for head-scratching headlines, such as this one from the left-leaning *High Country News*: "Can more oil extraction cut CO_2 emissions from power plants?"

Recent analyses suggest that EOR may sometimes help climate *a little*, but more often than not it doesn't help at all. That's because most EOR operations in the United States today use CO_2 *mined from underground*—with companies extracting the CO_2 they ultimately inject. Mining deposits of CO_2 that are stored safely away from the atmosphere and using them to push out oil could only happen in a country where the price of oil is high and carbon pollution is free.

Enhanced oil recovery is energy-intensive and pays for itself because of the additional oil produced. There's only so much EOR that's needed, however, and we need to store a lot more carbon dioxide than EOR will pay for. It's better than nothing for climate, but not by much. We need technologies that store all—or almost all—of the carbon dioxide released by burning fossil fuels in power plants. And pushing more oil out of the ground certainly won't help fight climate change.

ADM first considered an industrial-scale carbon capture and storage project from the coal plant that fuels the boilers in their operation. Exhaust from a power plant is typically about 10 percent carbon dioxide; thus, companies still have to capture and concentrate the CO_2 from a smokestack chemically, an additional energy-intensive and expensive step. As McDonald described it: "We were looking at capturing CO_2 from one of our coal-fired boilers and delivering it to an oil field for use in EOR. The project fell apart, though, when crude prices collapsed and we lost our oil partner."

Without an oil producer willing to pay ADM for their carbon dioxide, the project couldn't turn a profit. "The economics did not look good," Scott said. "It was going to be expensive to run and, of course, the capital costs to capture the carbon dioxide were substantial. Even if we received a lot of grant funding, it would have been very difficult to continue to operate the unit."

Here, then, is the rub for carbon capture and storage: it's expensive, and ADM couldn't do it unless someone like the government or an oil company paid them to do it. Alternatively, U.S. policy could require it, or at least require companies to pay a fee to pollute the atmosphere. Carbon capture and storage will never happen broadly—certainly not at the massive billion-ton scales needed—without either a policy mandate to end fossil pollution or a carbon price that makes it worth more to capture the pollution than to dump it into the air. If companies had to pay to pollute today, renewables would be even cheaper compared to fossils than they already are.

Cost, more than technology, is the limiting factor for most CCS projects today. No state between Nevada and Pennsylvania—including Illinois—had a price on carbon dioxide pollution when ADM was making its decision. Moreover, companies like ADM compete in a market where almost everyone—in the United States, at least—pollutes for free. Compared to free, doing *anything* to limit greenhouse gas pollution will always be more expensive.

So how did ADM pay for the sequestration project at their ethanol unit? They didn't—at least not fully. The federal government—and by extension, taxpayers like you and me—paid for most of it as a demonstration facility. Firstly, ADM dropped the more expensive goal of capturing carbon dioxide from their coal plant and used the waste stream from their ethanol unit instead, which is almost pure carbon dioxide. They only had to remove water from the waste gas, a relatively cheap step. ADM received $141 million in stimulus funds from the U.S. Department of Energy to kick off the project.

Beyond economic and technological constraints, social concerns can also limit deployment of CCS. Every CCS project faces the prospect of NUMBYism (used disparagingly for "not *under* my backyard"). Pumping millions of gallons of fluid underground raises concerns of CO_2 leakage and potential earthquakes. Several early CCS projects in Europe were canceled because of public concerns over leakage and safety. A project in Barendrecht, Netherlands, planned to store ten million tons of carbon dioxide from Shell's nearby Pernis refinery over a few decades. Residents complained that the project would endanger their town and reduce home prices. The municipal government opposed the project, and a local citizens group, "No to CO_2," formed quickly.

A postmortem by the Global CCS Institute titled "What Happened in Barendrecht?" cited shortcomings in communications as a cause for the project's failure. The authors recommended "creating mutual trust between stakeholders and commitment to each other and to the project," by "including all stakeholders in the project process at an early stage

and communicating about the project and its process to the community."
The project should also "be open and transparent to the participants, the
community and other stakeholders." Such recommendations for community engagement and openness are core tenets of the environmental
justice movement.

Resistance to CCS goes beyond safety, though. Opponents are angry
at fossil fuel companies for the damage their products have caused and
oppose anything that extends the industry's life. The idea of taxpayers
paying fossil fuel companies to capture and store their carbon pollution
is also anathema. Consequently, many green groups around the world
oppose CCS projects.

In 2019, 626 organizations, including Greenpeace USA, Physicians
for Social Responsibility, and the Sunrise Movement, sent a letter to the
U.S. Congress outlining their vision for a Green New Deal. The letter
rejected calls for carbon capture and storage, saying the groups would
vigorously oppose any legislation that "promotes corporate schemes
that place profits over community burdens and benefits, including carbon and emissions trading and offsets and carbon capture and storage."
The letter ended: "Fossil fuel companies should pay their fair share for
damages caused by climate change, rather than shifting those costs to
taxpayers."

ADM navigated a minefield of potential resistance as they sited their
CCS project in Decatur. Scott McDonald said, "Early in the process, we
conducted significant outreach to, first, local landowners around our facility, and then to the city council of Decatur, its mayor, and its city engineer. We talked to them about the project, what we were doing and what
the impact would be on our local economy." McDonald gave presenta-
tions to the City Planning Commission because they needed to rezone
some property to build an electrical substation. ADM also partnered
with Richland Community College to develop an educational program
that included one of the first CCS-related curricula.

I asked McDonald if there were any protests. "No," he said. "Part of the reason might have been because we're capturing biological rather than fossil carbon dioxide. That might have helped deter some NGOs from protesting the project. The local constituency was extremely supportive. We didn't really get any pushback." Decatur is something of a company town and the fact that ADM has been active there for more than a century probably helped also.

Alison Taylor weighed in as well: "I would have thought there might be more concern, because this is something novel. It might even have sounded like a massive industrial project that people wouldn't want to look at. When I finally saw it, I was surprised at the minimal footprint—how little you can see. So it's funny when people say they want to see it; there's really not much to see." There is piping to move the carbon dioxide from the ethanol plant to the point of injection, only a few miles or so away, and a couple of euphemistically named "Christmas trees" on site, the small pipes that plumb the CO_2 more than a mile belowground. There is no noise and no pollution at the point of injection.

Unlike Barendrecht, Decatur proved a winning location for both BECCS and the CCS that goes with it. But to store a billion tons of carbon dioxide a year, thousands of ADM-scale plants must be built, each project a minefield of potential opposition. The good news for successfully placing CCS operations in the United States is that many of the best potential storage sites are in relatively oil-and-gas-friendly states, such as Texas, Louisiana, Wyoming, and North Dakota.

Decisions about whether and where to site CCS facilities won't, and shouldn't, be made solely on technical and economic grounds, though— that is, where CCS can work best and what it will cost. Social concerns around the practice of CCS and other climate solutions are also important. I mentioned one earlier: that CCS facilities will prolong the life of fossil fuels. I support this concern—because replacing fossil fuels with low-carbon alternatives is the fastest path to a safe climate. Nevertheless, I believe we will need CCS in some form and in some places. Another social concern is whether CCS will prolong injustices for people living near polluting infrastructure who tend to be poorer and, often, people of color.

The words of environmental justice icon Catherine Coleman Flowers spring to mind (see Chapter 1). "One of the things that's concerning to people in the environmental justice community," she told me, "is that if the focus is on carbon capture, it doesn't get those dirty plants out of their communities. Such plants are more heavily located in communities that are marginalized, and that's not going to change if carbon capture is our focus. It only means that they'll still be in the Black communities and the brown communities and the poor communities. And if they have to bury some of that stuff, where are they going to bury it? It's generally 'not in my backyard' everywhere else so it ends up back in the same EJ [environmental justice] communities.

"There has to be balance in terms of where they're located," she said, adding that in fact something more radical than "balance" will be required: "Maybe the solution is to change technologies altogether. Maybe we'd have a different kind of manufacturing if polluting facilities were evenly distributed across affluent communities, instead of just in EJ communities."

10

Stoned

You can chase plumes of steam across Iceland's moss-pillowed landscape for days. After watching clouds mushroom from the Geldingadalir volcano near Reykjavík, I tack overland to a wine bottle–shaped steam stack marking the Efri-Reykir geothermal project in the Biskupstungur ("Bishop's tongues") district. I head to the Haukadalur Valley next—a favorite filming ground for *Game of Thrones*—to see Strokkur geyser swaddling people and hills in pleats of steam.

Iceland is electric-green—across its summer landscape and its power sector. It's one of the few countries in the world producing all of its electricity from renewable fuels—mostly hydro, geothermal, and wind power. Because of this abundant and cheap clean energy, Iceland is also a global leader in aluminum smelting and other heavy manufacturing (the world needs to produce aluminum somewhere, and ideally in a place with abundant green energy). It's also a destination for carbon capture and storage and other negative-emissions technologies that require zero-carbon energy. Iceland is an innovation hotbed, spawning new climate technologies.

I'm here to answer a question: Can we somehow convert atmospheric carbon dioxide directly into minerals within decades, instead of the millions of years it takes for nature to do so in our oceans?

A few hours after visiting the Geysir field in the Haukadalur Valley, I turn northeast off Highway 1 and drive down a red-and-black basalt ridge—a prostrate, petrified dragon whose nostrils bleed steam at the ridge's end—which marks the Hellisheiði geothermal plant. Steam plumes, from natural seeps and the plant's cooling towers, dot the landscape and mirror the cotton-ball clouds floating low in the sky nearby.

Hellisheiði is Iceland's second-largest power plant. The landscape surrounding it is plumbed like a green semiconductor chip with silver pipes—four to eight abreast—moving hot water and steam in parallel right-angle symmetry. I pull over and gently touch a pipe at the roadside to see if it's warm. It isn't, but it is surprisingly big—close in size to the tubes kids shoot down in enclosed water slides.

Energy companies in Iceland and elsewhere extract steam and heat from geothermal fields. Steam turns the turbine blades and rotor shafts of generators that make most of the world's electricity. In fact, making steam is *the* most energy-intensive step for fossil power plants, yielding billions of tons of carbon dioxide pollution yearly. Geothermal plants like Hellisheiði extract "free" steam heated by magma gurgling just below the Earth's surface, avoiding the cost and pollution of fossil fuel use.

But Iceland's Hellisheiði plant also eliminates carbon emissions in a second way. The plant sends hot water through insulated pipes to Reykjavík in a process called "district heating." This "free" hot water heats homes and buildings and eliminates the need for—and emissions from—gas furnaces and household water heaters. I wasn't able to relax in one of Iceland's abundant geothermal hot springs during my stay, but I was able to luxuriate in a warm shower powered solely by clean energy.

Geothermal power even has advantages compared with other clean renewables. It's a baseload fuel, dependable day after day, month after month, spring, summer, fall, and winter. There is no slow season for magma, the way there is for sunlight or wind. As a result, you don't need as much backup storage from batteries and other sources to navigate "off days" on the grid. Geothermal power is also far cleaner and cheaper than fossil power, just as wind, solar, and hydro power are.

Geothermal resources are abundant globally. Far from Iceland, the Ring of Fire that rims the Pacific Ocean with volcanoes and earthquakes yields geothermal hotspots for countries that include Japan, Indonesia, the Philippines, New Zealand, Russia, Mexico, Canada, and the United States, the world's leader in geothermal power generation. The world's biggest geothermal plant, "The Geysers," is located just seventy miles north of San Francisco. By quietly gathering geothermal steam from posh, wine-making Sonoma, Lake, and Mendocino counties, it provides enough electricity to power San Francisco.

Beyond chasing plumes of steam, I'm visiting the Hellisheiði power plant to see a pair of climate milestones. Hellisheiði isn't just a low-carbon powerhouse. It's the place where a company—Carbfix—is capturing carbon dioxide as a waste gas and storing it permanently underground. Carbfix mixes reservoir and chemical engineering with a splash of geology to realize its company slogan: "We turn CO_2 into stone."

I'm speaking with Dr. Edda Sif Pind Aradóttir, CEO of Carbfix. Dr. Aradóttir is a Viking dynamo in the fields of climate solutions, reservoir engineering, and geology. Before trying to save the world, she wanted to be a "real doctor." "I was going to be a medical doctor," she said, "but I didn't want to still be studying when I was forty. I decided to check out the renewable energy sector and haven't looked back." Aradóttir's list of degrees and accolades is longer than the Mid-Atlantic Ridge underlying Iceland.

The Hellisheiði power plant began by drilling some thirty wells a mile or two underground. The wells gather steam and hot water across the landscape and pipe it to the geothermal facility. By starting with "free" underground steam, the power plant emits only one-fiftieth of the carbon emissions of a fossil fuel plant for the same amount of energy. This geothermal energy still isn't completely carbon free, however, because some natural carbon dioxide bubbles up from underground with the water and steam. Before Carbfix came onto the scene, the Hellisheiði plant vented this carbon dioxide into the air as greenhouse gas pollution.

Typical carbon sequestration—if there is such a thing—can't work in Iceland. The dozens of industrial plants around the world currently sequestering carbon dioxide pollution pressurize the gas until it resembles a liquid, making it cheaper to pump underground. But this dense CO_2 is still buoyant—lighter than the groundwater it enters. It rises until a caprock or impermeable barrier blocks its way (the way a cork traps bubbles in a champagne bottle). Many places around the world have caprocks or other impermeable layers that have kept ancient gases trapped for millions of years; without them, most fossil gas production would be impossible.

In contrast, the rocks around Hellisheiði and across Iceland have been ripped and cracked by recent tectonic activity and can't hold gases or buoyant liquids tightly. As Aradóttir says, "We can't use a standard process because the bedrock in Iceland is young, porous, and fractured. We simply don't have good caprocks." She and her colleagues designed a new approach to overcome this limitation. Instead of pumping pure pressurized carbon dioxide gas underground—as other projects do—they dissolve the carbon dioxide in water and pump the sparkling water deep into the earth, where nature mineralizes it.

To do this, Carbfix's engineers designed an industrial-scale version of a kitchen appliance that uses a cylinder of carbon dioxide to make fizzy water. Pumped into the right rock, the dissolved carbon dioxide gas precipitates into carbonate minerals.

Volcanic basalt found across Iceland turns out to be one of the "right rocks" for mineralizing carbon. Basalt is one of the most common rocks on Earth and it underlies much of Iceland. Basalt contains lots of the calcium-, magnesium-, and iron-rich minerals that can help form carbonates and other carbon-rich rocks, such as limestone (mentioned in Chapter 8). You can also precipitate carbon to form "magnesium limestones" and calcium-magnesium hybrids called "dolomites," like the classic mountain range of the Alps.

I've seen this process firsthand while studying water movement underground in caves. Stalagmites and stalactites are made of the calcium carbonate precipitated from carbon-rich drips of cave water.

Basalt is often gray or gray-brown. I pick up a rock core that came from under the Hellisheiði carbon storage facility. It's sandpaper-rough and resembles Rocky Road ice cream, except the chocolate ice cream is gray and the marshmallow nuggets are pockets of new white carbonate minerals. This mineralization is permanent. Aradóttir explains, "With our process, there is no buoyancy and no risk of CO_2 reaching the atmosphere."

To confirm mineralization, Aradóttir and her colleagues used experiments to track the fate of carbon dioxide they pumped underground. They injected tiny amounts of carbon dioxide labeled with radioactive carbon and measured changes in its concentration between the injection point and a monitoring well a football field's distance away. More than 95 percent of the carbon dioxide injected below the Carbfix site turned to carbonate minerals *in less than two years.*

That finding shocked everyone. Prior to Aradóttir's work, most geologists believed mineralization would take centuries or more to occur after injection. A year and a half into Carbfix's test, the submersible pump in the injection well broke when calcite ($CaCO_3$) coated and clogged it—a clear indication that mineralization was occurring. (Calcite is the limestone "scale" that builds up on hot tubs, bathtubs, and tea kettles in calcium-rich water.)

I ask Aradóttir why mineralization didn't limit injections into the well, clogging the closest pores and increasing the energy needed to pump the carbonated water farther away. "If all the pores near the injection well filled up first, that would be bad," she says. But that didn't happen because the fizzy water is acidic and dissolves some of the rock near the injection well. "Our ability to inject water has actually improved since 2014," she adds.

I ask her what the biggest surprise of the project was—good or bad. "We've had loads of challenges," Aradóttir says, "especially bringing an idea from paper to an actual, operating industrial process. What surprised me the most is how smoothly it's working. It's a simple and beautiful process. Today, it's automatic. During the hassle of the pilot stage I didn't expect that outcome. It makes scaling up so much easier." She adds, "It's not rocket science, it's rock science."

Headwinds slow me as I walk the Hellisheiði site to see a handful of diminutive geodesic domes dotting the landscape. These silver igloos hold the wells where Carbfix injects carbonated water underground. Leaning into the stiff Icelandic wind, I wonder what headwinds the company faces in scaling its technologies elsewhere, including the possibility of inducing earthquakes or contaminating groundwater.

Pumping a large quantity of any fluid underground can cause tremors or earthquakes that damage homes and buildings, as has happened with fracking in the Netherlands and with wastewater disposal from fracking in Oklahoma. Starting around 2014, Oklahoma suddenly had more earthquakes than California—a thousand magnitude 3 earthquakes in 2015 alone, compared to just two a year on average from 1979 through 1998. A magnitude 5.7 earthquake near Prague, Oklahoma, attributed to the injection of wastewater from oil and gas extraction, cracked windows and masonry in local homes and collapsed a turret at St. Gregory's University.

Pumping too much water too fast (and sometimes into the wrong layers) caused most of the problems. In time, new state regulations and better practices from disposal companies reduced the number of earthquakes substantially. Oklahoma implemented a "traffic light" system, assigning "yellow light" status to some disposal wells (reducing their allowed injection rates and pressures) and "red light" status to others, shutting them entirely. By 2019 and 2020, the number of magnitude 3 earthquakes in Oklahoma had dropped to about 150 a year, far above prefracking levels but well below the 2014 peak.

"We haven't seen earthquakes with our CO_2 injections," Aradóttir says, "but there were some tremors a decade ago when the power plant first began pumping water up and down. Now we use a traffic-light system to minimize risk, too."

Aradóttir is quick to highlight the differences between their carbon storage in Iceland and more common approaches for carbon sequestration and wastewater disposal elsewhere. "Our CO_2 injection wells are only five hundred to eight hundred meters deep," she says. "We can thus inject at much lower pressures than for typical carbon storage wells that are much deeper. And we have monitoring networks, of course, to measure any warning signs."

Groundwater contamination can be another concern when pumping fluids with chemicals underground, as we do when fracking. But because Carbfix adds only carbon dioxide to their water, public concerns over contaminating groundwater are relatively low in Iceland. In fact, "the water is purer after we mineralize carbon dioxide than before," Aradóttir says. "Our process *reduces* concentrations of toxic heavy metals found naturally in groundwater."

Other factors could limit the use of Carbfix's technology, including a lack of fresh water in drier regions of the world. To explore alternatives, the company is exploring using seawater for pumping carbon dioxide underground. If Carbfix succeeds, as results to date suggest they will, water demand should be less of an issue when they scale their technology to drier areas.

Another impediment to scaling Carbfix's approach and to other technologies is getting carbon dioxide to the storage sites. "There is a lot of ongoing work, at least here in Europe, to develop CO_2 transport networks," Aradóttir notes. "These networks would gather CO_2 from different parts of the European continent and bring it to larger storage hubs. I don't know what the situation is in the United States to develop similar CO_2 transport networks," she says. In fact, today the United States has only about five thousand miles of CO_2 pipelines, used mostly to transport carbon dioxide to oil fields for enhanced oil recovery, as discussed in the last chapter.

Because an undersea CO_2 pipeline covering thousands of miles from continental Europe to Iceland would be prohibitively expensive, Carbfix is seeking other ways to transport carbon dioxide. They recently received a €117 million grant from the EU's Innovation Fund to build the "Coda Terminal," a facility in southwest Iceland designed to store three million tons of carbon dioxide a year. The terminal will receive ships from northern Europe and store their carbon dioxide in basalt formations nearby.

Beyond bringing carbon dioxide to Iceland, Carbfix hopes to export its technology to other countries. One recent project funded by the European Union applies Carbfix's Iceland method to geothermal plants in three new countries: Italy, Turkey, and Germany. In Italy, Carbfix is drilling wells to generate geothermal power and trying to reinject all waste gases, including carbon dioxide, without releasing any into the air. They're also testing their process in additional rock types beyond igneous basalt. "Local geology is important," Aradóttir says. "To keep costs down, it helps to be located as close as possible to storage formations."

In fact, cost, more than technological feasibility, remains the bigger barrier to climate action, including large-scale carbon capture and storage. "Cost is holding climate action back," Aradóttir says. "Everyone wants to reduce their emissions, but not many want to pay for it. They want someone else to pay. This has been the situation for too long." Carbfix's process costs less than $25 per metric ton of carbon dioxide stored, a price that is already competitive with carbon markets in Europe's Emissions Trading System.

Sometimes green solutions like those at Hellisheiði succeed because they provide extra benefits that gain public support. Since arriving in

Iceland—near the airport, driving near Reykjavík, visiting the geysers, and even here at Hellisheiði—I've sometimes sniffed rotten-egg sulfur that wafts from hot springs and geothermal fields. Along with carbon dioxide, the Hellisheiði geothermal plant used to vent this "sewer gas" (smelly hydrogen sulfide or H_2S) that rose to the surface with the water and steam. On most days the smell dissipated quickly in the wind, but occasionally sulfur pollution in stagnant air drifted to Reykjavík, only twenty or so miles away.

"Before our carbon-control project," Aradóttir says, "hydrogen sulfide emissions from the geothermal plant sometimes stank in the city. Air pollution was a big thing. However, we knew we could co-capture and mineralize hydrogen sulfide as an added benefit to mineralizing carbon for storage. We've reduced hydrogen sulfide emissions from the power plant by 75 percent. You almost can't smell it anymore—something the public is very happy with."

There's still more work to be done at Hellisheiði. Currently, Carbfix is only capturing one-third of the plant's carbon dioxide emissions, but they are investing in new equipment to store the rest. Only when they can capture *all* of the emitted carbon dioxide will the plant be truly carbon neutral, emitting zero carbon dioxide for each additional unit of power produced.

"Turning carbon into stone" is a milestone, but carbon-neutral technologies alone won't be enough to keep global temperatures at safe levels. We need carbon-negative solutions, too—technologies that remove carbon dioxide from the air and lock it permanently away. And another revolution is happening at Hellisheiði—just a stone's throw from where I stand.

To run greenhouse gas emissions backward, Carbfix is teaming up with the Swiss company Climeworks, based in Zurich. Created by two mechanical engineering students in response to the shocking retreat of glaciers in the Alps, Climeworks is pairing its unique carbon capture chemistry with Carbfix's technology to pump fizzy carbonated water underground to mineralize the carbon.

Christoph Beuttler, manager of carbon dioxide removal and chief climate policy officer at Climeworks, told me that Carbfix's operation is approaching being carbon "neutral"—generating power while contributing no greenhouse gases to the atmosphere. The goal of Climeworks+Carbfix is to be carbon "negative" eventually, removing carbon dioxide from the air and storing it permanently underground.

"We deal every day with how hard it is to remove carbon dioxide from the atmosphere. I call it chasing 420 parts per million," Beuttler said, referencing the concentration of carbon dioxide in the atmosphere when we spoke. "It will always be more expensive than renewable energy. Continuing to burn fossil fuels today and removing them later with direct-air capture [DAC] makes zero economic sense." Doing so is energy-intensive and expensive—the cost is at least $500 per ton of carbon dioxide captured from the air for most industrial companies currently.

The Hellisheiði process is much cheaper because they don't have to capture CO_2 from air and they only have to pump it into relatively shallow layers. Carbon removal companies need to reduce prices to $100 per ton or less for them to make a substantive contribution to climate solutions. Achieving this price drop may take decades, just as it did for solar power to become cheaper than fossil power.

Still, if we want to alleviate some of the suffering we're already experiencing worldwide—with additional suffering to come—we'll need to remove greenhouse gases from the atmosphere. The same holds true for stabilizing the Earth's temperature at no more than 1.5 or 2°C above preindustrial levels—if we keep emitting more greenhouse gases than the atmosphere can bear for those thresholds. In fact, direct-air capture of carbon dioxide is already included in most energy and climate scenarios in which the Earth's average global surface temperature increase stays below 2°C, and essentially *all* scenarios for the more ambitious

1.5°C target. Countries are banking on direct-air capture to work even before it's widely available, and that's dangerous while we keep releasing more carbon dioxide and methane pollution into the air.

From today onward, every additional ton of carbon dioxide we emit will need to be removed by someone, somewhere, sometime, and should come with a price of carbon removal attached to it, far higher than the current price in Europe and elsewhere. Few people discuss this need because it would make fossil fuels impossibly expensive compared to other fuels today. By ignoring this future cost, we're foisting it onto young people and future generations. As Christoph Beuttler put it, "We have to go down the emissions reduction path as quickly as we can. We see DAC [direct-air capture] as being there for unavoidable emissions you can't get rid of, and to enter net negative territory. The big question is who pays for it."

Well, Microsoft, for starters. Climeworks already has commitments as part of Microsoft's carbon removal portfolio, to help the company reach negative emissions by 2030 and remove the company's historical emissions by 2050. Climeworks also has a successful subscription program where tens of thousands of individual "Pioneers" worldwide have committed to pay to remove hundreds of pounds of carbon dioxide from the air each year, at a typical subscription cost of between U.S.$10 and $50 a month. Why would someone do this? According to Beuttler, "Some like the technology. Some realize if they don't help us scale our technology, it will never grow. Some want to offset their personal footprint."

Climeworks' carbon capture arrays at Hellisheiði look like basalt-gray container cars stacked on a train or cargo ship. An individual "car" has six cubical units nested side to side. Each unit sports a pair of jetlike fans, one for taking air with carbon dioxide into the unit and the other for pushing mostly carbon-free air back out. Once inside the unit, the carbon dioxide gas reacts with an amine, a nitrogen-based compound. When the filter is saturated or "full," Climeworks closes the unit and raises its temperature to the boiling point of water, using renewable energy from the geothermal plant. The carbon dioxide releases and can be pumped underground in Carbfix's carbonated water, to be stored permanently at Hellisheiði.

Climeworks opened the first commercial air-capture plant for CO_2 in Switzerland in 2017. The plant captured about nine hundred metric tons of carbon dioxide a year. The biggest surprise was "just accomplishing it," Beuttler said. "Before we built the big plants in 2017, I think a lot of people said this is a lab experiment, this can't be done large scale, or this can't be done—full stop. Doing it changed a lot of people's minds."

The newer Hellisheiði plant captures nearly five times as much as that first facility—still a drop in the bucket compared to the billions of tons of CO_2 emitted to the air annually from fossil fuel pollution. Climeworks' goal is to have, within this decade, individual facilities each storing a million tons of carbon dioxide a year.

Christoph Beuttler listed some barriers to their success: "The biggest challenge is to scale our technology in time to be climate relevant. There are technical challenges. Do we have enough resources and materials, and can we produce them in an environmentally friendly way? Will there be enough green energy for our process?" He mentioned social factors, as well: "There are challenges with perception—where moral hazard might play a role," he said.

"Moral hazard" is the possibility that carbon removal in the future will reduce momentum for emissions cuts today. Carbon removal could be—and may already be—used as a delaying tactic. My Global Carbon Project colleague Glen Peters coauthored a 2016 paper in *Science* that included this passage: "Negative-emission technologies are not an insurance policy, but rather an unjust and high-stakes gamble. There is a real risk they will be unable to deliver on the scale of their promise." The paper concluded: "The mitigation agenda should proceed on the premise that they will not work at scale."

I used to be more concerned about moral hazards in the past than I am today. We have continued polluting the atmosphere with greenhouse gases for decades, despite conclusive evidence for, and awareness of, their dangers and consequences. In my view, human behavior remains the greatest hazard—moral or otherwise.

Christoph Beuttler gave another justification for pursuing carbon removal from the atmosphere: "Zero emissions is impossible. There's only net zero that's possible because you can't get completely to zero emissions in most sectors. It's a monumental task," he said. "This is our last chance."

He was also optimistic about recent progress in climate policy. "Three or four years ago we almost had no net-zero commitments ratified by countries. We didn't have a single country on this planet that had a position on carbon removal. And now so many countries have joined net-zero pledges: China, which emits 28 percent of global fossil carbon emissions, pledged carbon neutrality by 2060, and in the United States carbon removal is now almost mainstream, at least in climate policy circles. The progress is insane. Three years ago nobody would have thought that. It's still not enough, obviously, but it makes me hopeful. Insane is not the best word. It's just crazy good."

11

RePeat

Vedetään nuottaa
Saadaan kaloja
Vedetään nuottaa
Saadaan kaloja . . .

'm barefoot in the twilight waters of Lake Ylinen singing a Finnish fishing song ("Let's pull the seine, / Many fish to gain"). Rope puddles around me as I line in hundreds of feet hand over hand. "Slow down, slow down," my host and seining song-master, Tero Mustonen, says. "No more than four meters per minute." Simple enough, except for his prompts to "keep the net moving or all fish will escape" when I pause unconsciously to watch an Arctic loon or tern fishing nearby.

I wish drawing carbon dioxide and methane out of the air were as simple as drawing in fish with a net. Nature-based solutions nevertheless provide some of the cheapest ways to mend the atmosphere. As cofounder and president of the Snowchange Cooperative, Tero Mustonen, with his partners, is stitching damaged ecosystems back together. They're particularly interested in rewilding ecosystems ravaged by peat mining—such as the Linnunsuo wetland I'm visiting—restoring biodiversity and stemming losses of greenhouse gases. "It's critical to stop

144

soil-based carbon emissions from degraded peatlands," he says, "and to restore their power to absorb carbon."

"Tero" carries a slight roll of the "r" like the light waves rippling around my legs in the lake. "Mustonen" or "*Mus*-toe-nen" recalls the black-on-yellow "Moose Crossing" signs posted along every Finnish road I've driven. Round-faced with glasses, Mustonen is wearing a coral polar fleece jacket and a T-shirt with the Snowchange logo on it: a prominent salmon over a stylized snowflake with each of its six points representing a different indigenous people of the north.

"Snow unites all northern peoples and so do salmon," Mustonen says, explaining the logo's meaning. "This is the part of the world where the boreal, the high arctic, and tundra occur, where human societies and nature depend on proper ice and snow. We're called Snowchange because snow links us, and we believe change must be done urgently." He knows snow and ice because commercial fishermen in northern Karelia—like him—spend winters on the ice passing nets through holes into the lake below.

Snowchange is as multifaceted as Tero Mustonen. As a global network of indigenous cultures, including Sámi, Chukchi, Inuit, Inuvialuit, Inupiaq, Icelandic, and other tribes, Snowchange documents indigenous observations of climate change in northern Canada, Scandinavia, Russia, and the United States. It is also a scientific organization that works with the Arctic Council and the Intergovernmental Panel on Climate Change, the United Nations body investigating climate change—for which Mustonen is a lead author on issues of climate, biodiversity, and resilience. Snowchange is a cultural treasury that holds traditional knowledge sacred, including fishing and hunting, handicrafts, stories, and other elements of forest culture.

Snowchange also rewilds habitats. Rewilding is practiced most commonly in Europe, where it is often described as a more "hands-off" or "progressive" approach to conservation. As the group Rewilding Europe defines it, "It's about letting nature take care of itself, enabling natural processes to shape land and sea, repair damaged ecosystems and restore degraded landscapes." The group suggests: "We can give [nature] a helping hand by creating the right conditions—by removing dikes and dams to free up rivers, by reducing active management of wildlife populations, by allowing natural forest regeneration, and by reintroducing species that have disappeared as a result of man's actions. Then we should step back and let nature manage itself."

For Snowchange, rewilding diverges from traditional restoration in placing less emphasis on recreating what *used to be* and more effort on creating habitats that will last into the future. Providing long-term natural and cultural benefits in the face of climate change is a particularly important goal for the group.

As thoughts about rewilding float by like leaves on the water, Mustonen barks, "Keep the net moving!"

When the seine finally arrives, we gather a modest harvest of vendace, red-eyed roach, whitefish, and fishery-raised salmon, evident from the distinctive fin clipped near the tail. Together with local students and Snowchange staff, we circle around a crackling fire on the shore for warmth (and to let the smoke drive the mosquitoes away). Then, using paper towels as plates, we share grilled fish and stories.

Mustonen likes to say, "Finns can be silent in many languages simultaneously." Fortunately, he isn't. Discussing the importance of fishing personally and to the region, he says, "I grew up fishing. . . . We had a household full of fish traps and gill net equipment, and my family sold fish and crayfish commercially as supplementary income. Regionally, fishing is probably the longest thing Finno-Ugric peoples have done. The world's oldest net ever found by archaeologists came from here and was made more than ten thousand years ago." Gesturing to the northern sky, he adds, "In Finnish, the Big Dipper is called *Otava*—a type of seine or salmon net."

Five herons beat by in the twilight, their screechy croaks breaking the silence. "Gray herons." Mustonen says, "A sign of climate change." Like other species we've seen this week, including the European hare expanding into the boreal taiga, the gray heron is extending its range northward across Finland and Sweden and is now a regular visitor even in Iceland and Greenland.

Climate change motivated Mustonen and others to found Snowchange. "The north needs snow and ice for life," he says. "Snow protects plant life and, in some cases, the birds and animals in the deepest, frostiest months. It enables transportation, including travel on the lakes for winter fishing, where poor ice is now a safety issue.

"Climate change is hitting us harder here in the north than anywhere else, except perhaps in the Pacific. It's very disturbing because whole ecosystems are changing, especially when you have a lot of rain in winter. The cryosphere links life and cultures in codependency. If you take away the most important part—the cold—it cascades across ecosystems and cultures."

Mustonen mentions the close relationship between winter fishermen, like himself, and the land: "People involved in fisheries or on the lakes are in a way guardians of the lakes because they carefully observe daily weather quality and ice conditions, including when the ice comes and goes. And by 1996 and 1997 they jumped to the conclusion that winters had changed into something completely different. There was regular rain coming in February, 7 or 8°C above zero. I recall in December 1996, walking around a small lake and it was plus 10°C and really misty. It was almost like October weather except that it was supposed to be minus twenty and, of course, completely frozen."

"One year in particular is a benchmark in northern Finland," he says. "Nineteen eighty-seven is the year known here as the 'farewell to winters.' It was the last proper winter with temperatures below -30°C for weeks.

"Building on common experiences and shared observation of change, we came together two decades ago [around 1999] and decided, 'Change can happen in other directions.' We don't have to just be victims of climate change; we can set up an organization and network to respond to it. And at that time, we also had some contacts with the Sámi people, the reindeer herders, including a reindeer herder named Stefan Mikaelsson who was the vice president of the Sámi Council. I also had an Inuit friend and Nunavut colleagues from Canada who were interested."

Mustonen musters disparate peoples the way a musher whistles sled dogs. "Sometimes I feel like a person of two worlds," he says. "I'm leading an organization that supports and works with indigenous communities, and on the other hand, I'm sitting on the IPCC and other scientific bodies trying to be the interpreter of both worlds. The science world is more like an arrow. It has an analytical mandate or framing and it looks at one specific thing. For indigenous peoples and local communities, however, the flow of information isn't just on the left side of the brain—it's your hands, your feelings, your emotional attachment to land and change. When you think of indigenous engagement with their homelands, the emphasis is on holistic engagement."

The notion that our lands are stable and will outlast us lodges deeply in our psyches. Arctic lands, however, are cratering, oozing, slumping, sloshing, stretching, seeping, sinking, and slipping away before our eyes or—more accurately—the eyes of the northern peoples who experience such changes daily. "Reindeer herders see their landscapes sinking a meter a year," Mustonen says. "Methane is being released from soils in Siberia and in the East Siberian Sea, where bubbles emerge. And then there are the methane burst craters, especially in Yamal but also in Yakutia, in Eastern Siberia, that have accelerated since 2014. Reindeer herders have been the first to observe these craters associated with methane bursts." I'd read about such burst craters that appear unpredictably in Arctic permafrost landscapes, leaving pockmarks fifty meters deep and twenty meters wide. Their cause is mysterious.

Mustonen gestures vaguely to the north. "What I and others have most observed in the high Arctic tundra is the massive permafrost melt. I traveled, for example, twelve hundred kilometers in a small, open boat on the East Siberian Sea coast, from the Kolyma Delta all the way to Chaigurgino, a small seasonal fishing camp on the Bolshoi Chutkotskaya River. Kilometer after kilometer, I saw collapsing permafrost walls. I heard the *sloosh* of the coast slumping into the sea," he says, marking the downward slip of the coast with his hands. "There was a weird smell, too, as if you had a gas lamp running without a spark.

"Before that trip, I already had concerns about climate change. But this was my first long trip in the summer to these regions with the Chukchi reindeer herders.

"I had no words," he says, about seeing ecosystems collapse before him for hundreds of miles. "I decided, 'I must press every button in my power to convey how bad this is.'"

Beyond documenting how bad things are getting, Tero Mustonen and Snowchange are actively repairing ecosystems to combat climate change. I met him and his students at the peatland rewilding site of Linnunsuo. It's a wetland near the village of Selkie, in the belly of eastern Finland that bulges into Russia. Today, Linnunsuo is one of Finland's premier hotspots for wading birds and waterfowl. During spring and fall migration, more than two hundred bird species can be found here.

A decade ago, though, the Linnunsuo wetland was a Mordor-like slag heap, the legacy of industrial peat mined for energy. Only two bird species used the site. Mustonen and Snowchange have spent a decade returning Linnunsuo to something functional—a "novel ecosystem" rather than a faithful re-creation.

As Mustonen and I walk a trail around the Linnunsuo wetland, our shoes crunch on the gravel path. We pass dark, rectangular scars where industrial diggers clawed peat from the ground. "This is the first rewilding site in Finland," he tells me. He dismisses the idea of restoring the site to some preindustrial ideal. "This was a moonscape. There is no such thing as restoration here; these recovering ecosystems will never be exactly the way they were in the 1800s. They will be living in a warming planet no matter how successful we are. Nobody has a map of how boreal and Arctic systems will respond to changes in precipitation, moisture, drought, and the loss of snow and ice.

"Wait," Mustonen tells me, placing his arm in front of me. We pause briefly on the trail crossing the wetland. He pulls out his camera to photograph a yellow-green-tinged dragonfly—perhaps a female green darter—now perched on my black fleece jacket. He mentions how dragonfly larvae need good water quality to survive, an indication that the landscape is healing. I snap my own photo when the dragonfly darts from my shoulder to the ground.

It lands on a mix of white pebbles and larger black and brown stones near our feet. The white pebbles are limestone, a calcium carbonate mineral added to soils and water to reduce acidity. A water treatment company gave Snowchange piles of minerals to spread throughout the wetland, to reduce the dangerous pH levels left after peat mining.

Earlier, we'd stopped at a red hut next to the water. Bending over to enter, I see a woodstove, cot, and blankets, a well-used pH meter to test the water, a half-liter bottle of Koskenkorva vodka marked with "pH" in faded Sharpie—most likely a calibration solution for the pH meter—and some aerial maps of the site's wetlands tacked to the walls. It is a cozy retreat for fieldwork.

Mustonen points to the maps and recounts the history of Linnunsuo wetland. "Between 1984 and 2010 one of Finland's state-owned enterprises was digging peat here for energy production. Then in 2010 there was an acidic discharge from this ditch to the river"—pointing to the small waterway just outside our door—"after iron sulfate in the soils became exposed to air. We had 2.77 pH acidic waters leaving here that killed all life in the Jukajoki River downstream. The event was not detected by the company but by our fishermen. We issued a dire warning to the company and authorities that there has been a big pollution event and that the whole river was dead. Then one year later at exactly the same time—June—there was a big warm spell followed by a huge thunderstorm that led to another release of acidic water of pH 2.7 or 2.8. All the fish died again. The river had just started to recover from the first death event."

Tero Mustonen and others said, "Enough." Beyond being Snowchange's president, he is also the leader of Selkie village, a boreal settlement at least five hundred years old. "As head of the community, I informed the company that it's time for them to pull out. We took them to court. There was lots of media hassle and attention. And for the first time in Finnish history, the company pulled out of a village. We won a big victory.

"As a result of the attention," he says, "the company made the three first wetland pools to control the acidic waters. The only way to do that was to drown the soils exposed from mining in a low-oxygen environment. However, the by-product was that it became one of the largest wetlands in Finland."

Fast-forwarding to 2017 after the site had become an important bird habitat, the company began talks to see if Snowchange would be interested in owning and maintaining the site. "After discussing the risks and benefits, we decided to adopt the wetland and try to fix the whole system," Mustonen says.

What fascinated me in hearing about this process was that they didn't try to return the site to a perfect peatland, as traditional restoration would have attempted. They were making do—and trying to make better—with something else, a novel wetland ecosystem that is now prime stopover habitat for migrating birds. Both the local landscape and the local people are adapting to climate change.

"Perhaps landscapes have the right to decide on their own terms which species come back and what food chains form," Mustonen says, as we approach the end of the loop trail. "That's the magic of rewilding compared to ecosystem restoration. In conservation biology, every little weed is nipped away along with every single species not wanted by humans. They want to create an artificial replica of something from the past. But we are never going back—in our lifetimes, at least—to something that was here in the 1950s or earlier. This is why I think 'rewilding' and 'novel ecosystems' are more appropriate terms."

We end our walk at a large interpretive sign for the Linnunsuo wetland. "Even this map doesn't show all the pools here now," he says. "There are nine different wetlands. And it's different from the original three the company made. This has become home for two hundred bird species, including extremely rare waders like the red sandpiper and long-billed dowitcher, and other birds that have benefitted from the wetland, such as the northern pintail and wood sandpiper.

"We are witnessing the creation or emergence of novel ecosystems and novel landscapes. They were pristine originally and then destroyed by human use. And now in a very exciting and profoundly transformative way, these rewilded landscapes are re-emerging as something new. They are fierce and they are stout and they are excited to be alive. And it's amazing to be standing in them and witness a hundred thousand geese landing in front of you in an autumn spiral."

As we reach the end of the loop trail, Tero Mustonen scans the wetland and asks, "Now, where did the boss go?" He's looking for his partner and fellow Snowchange leader, Kaisu Mustonen. Kaisu is a geographer who heads Snowchange's global biodiversity program and supervises the greenhouse gas measurements here at Linnunsuo. She's also worked for years in the Sámi territories in northern Scandinavia and with Siberian indigenous communities in Russia.

Tero spots Kaisu in the distance, off-trail with three or four other researchers. A northern harrier banks past us toward them, its wings held in a shallow V. Long-tailed with a distinctive white rump patch, it's hunting low over the wetland, arcing like a pendulum—back and forth, back and forth—a hypnotist leading us to Kaisu.

So far, Tero and I have discussed the benefits of rewilding at Linnunsuo for diversity and for species recovery. "We even have a resident wolverine here, and otters and foxes and other predators," he says. We'd also discussed the need to protect people, fish, and other species downstream from dangerous acidic releases. What we hadn't talked about yet was greenhouse gas emissions—how much can rewilding reverse the damage from destructive peat mining and return the site to removing greenhouse gases from the air?

As we approach, Kaisu sets down the LI-COR methane and carbon dioxide analyzer. We met at dinner a few nights before along with the graduate students from the University of Joensuu here today: Alicia, Antoine, and James. Alicia is carrying a gray hard-case backpack that holds a different intrument; armfuls of analyzer sampling tubes dangle over her left shoulder octopus-like to keep them from dragging in the mud. The other students are carrying additional analyzers and chambers as we cross a landscape pocked by peat mining. I stop at a scar and grab a handful of degraded soil. It's black and organic. Exposed to the air, as here, it's quickly decomposed by microbes releasing carbon dioxide as they go. To do nothing at Linnunsuo would mean that almost all of the remaining peat and soil carbon accumulated over centuries of plant growth would flood the atmosphere.

Kaisu and her team at Linnunsuo have been measuring methane and carbon dioxide emissions for more than a year along transects across the wetland. Wetlands are some of the most productive habitats on Earth, removing carbon dioxide from the air as their plants grow. However, flooding in peatlands and wetlands typically releases methane. That's because microbes generate methane in low-oxygen, waterlogged environments, including wetlands, rice paddies, and the sloshy stomachs of cows, which function as "walking wetlands." Thus, managing greenhouse gas emissions is a tradeoff among gases, balancing carbon dioxide uptake and methane loss. The cooling from the carbon dioxide stored as peat eventually outweighs the warming from shorter-lived methane, though it sometimes takes decades or even centuries of recovery for this to balance out.

Previous researchers have already shown that rewetting damaged peatlands can produce immediate climate benefits in some cases. One team titled their recent paper "Prompt Rewetting of Drained Peatlands Reduces Climate Warming Despite Methane Emissions." Other multi-year measurements in restored peatlands suggest it can take decades for the carbon uptake of plant growth to overcome the methane released. That's particularly true in warmer regions, where microbes grow relatively quickly and emit methane faster—but less true in colder regions, such as here at Linnunsuo, where cold temperatures slow microbes down.

Kaisu's team is quantifying Linnunsuo's carbon uptake and methane emissions. Alicia places a clear plastic chamber about the diameter of a dinner plate down first in a relatively dry grassy area and then in an area of black and degraded peat soil. The analyzer's fan whirs. "We measure for two minutes at each point," she says. The methane emissions are small in both cases. However the analyzer shows that the grassy area is taking up carbon dioxide, whereas the degraded black peat is losing carbon quickly.

A few minutes later, graduate student Antoine places his chamber over a wet, mucky area and the analyzer (and Antoine) buzzes. "It's nasty," he says. "Oh, my god. There is a lot going on here." He points to the analyzer's screen, which shows a large spike. "Here, you can see the methane went really crazy." He cautioned, though, against making too much of a single hotspot—emphasizing that they need more time to characterize the landscape fully. "You need to work across warm and cold seasons and over a patchy landscape of low and high methane emissions to characterize what's happening overall at the site."

Antoine, Alicia, and Kaisu aren't certain whether the rewilding at

Linnunsuo is helping the climate yet, but they know, through their own research and the work of others, that it will eventually. "By next year we should have comprehensive results for the long term balance of methane and carbon dioxide," Kaisu says.

When I ask Antoine how he feels about the decision to rewild the site, he says, "What was the alternative? It was a moonscape losing carbon. And we need to weigh more than just the climate benefits, including biodiversity."

Tero acknowledged the same tradeoff earlier while we walked: "The most important action to take on a degraded peatland is to stop soil-based emissions." Referring just to the Linnunsuo site, he said, "It's somewhere between four hundred thousand and six hundred thousand kilos of CO_2 emissions per year we're able to stop through our rewilding." A quick conversion reminds me that's like taking a hundred thousand U.S. cars a year off the roads.

As to the possibility that methane release may outweigh the carbon uptake for a time? "We accept a temporary increase in methane emissions," Tero said, "given the benefits for birds and biodiversity, the lower CO_2 emissions, protection from acid releases, and more.

"To turn this back into a breeding peatland, the so-called lungs of the northern world, will take decades. The sphagnum is starting to come back, but these are large pools, and it will take years for the plants to recover completely."

Antoine bends over and strokes the lush, green sphagnum—the group of tiny feathery mosses that hold water and often form peat—emerging in the damp at our feet. To my eye, they look like mini palm trees. "Sphagnum is the superhero of northern boreal lands," he says, "the source of most peat storage. And it's recovering before our eyes."

After Kaisu and the team finish their methane and carbon dioxide measurements, we eat lunch in an open-walled hut—with cut firewood stacked neatly beneath wooden benches and a small fire ring to warm us and cook over. Kaisu concludes, "Linnunsuo is recovering amazingly well. Nobody expected the birds to return so quickly." Still, she laments the damage that rewilding itself can require. "We had to rip the land apart again to rewild it, using heavy machinery—digging and scraping to create the dikes—to control drainage in the restored ponds and to de-acidify the water in the wetland. It was ugly at first and it hasn't been easy. We still have to add lime to fight the acidity, and we'll have to help the system recover for some time before it leaves its emergency state.

"As the years pass, I hope we can let the wetland do its own thing. The land is in a state of shock. This peatland took ten thousand years to form and twenty-five years to destroy."

12

X-Methane

There's more than one way to destroy a peatland. You can drain it and tear out the peat mechanically, as happened at Linnunsuo, or you can alter the environment in ways that stop peatlands from persisting, releasing thousands of years' worth of stored carbon. Climate change is doing just that in places.

Northern peatlands form in low-oxygen, waterlogged environments where microbial decomposition can't keep up with each year's new plant growth. As a result, dead plant material builds up in the soil over centuries. Work by scientist Gustaf Hugelius and my lab estimates that northern peatlands contain more than half the current atmosphere's mass of carbon as peat—four hundred billion tons' worth.

The fate of that carbon hangs in the balance. Thousands of years' worth of accumulated peat carbon may reach the atmosphere as carbon dioxide either through peat fires as the permafrost thaws or through microbial decomposition as Arctic and boreal soils warm. Alternatively, and depending on how swampy the thawed peatland is, much of the carbon may be released by microbes as methane. Either scenario is disastrous for climate, but the methane scenario is worse—and we'll likely get some of both. Atmospheric carbon dioxide concentrations could jump one quarter over today's level and so could methane's—depending on how fast the release comes.

Moreover, no technology exists today that allows us to slow this methane release once it starts—either in the Arctic or in tropical ecosystems like the Amazon. We need to develop technologies that allow us to destroy methane after it enters the atmosphere.

Such methane-removal technologies could be an insurance policy in the spirit that George Shultz expressed in Chapter 7. I've been working for years to build a community of researchers around methane removal, studying new technologies and outlining benchmarks for progress. We have our work cut out for us.

To start with, methane is two hundred times less abundant in the atmosphere than carbon dioxide. That makes it harder to isolate, a classic needle-in-a-haystack problem. Methane's simple pyramid structure and strong carbon-hydrogen bonds also make it more difficult to crack open than the straight carbon dioxide molecule, except at very high temperatures.

Methane removal has some advantages, though. For one, you don't need to capture it and pump it underground as you do for carbon dioxide. If you can oxidize methane using catalysts or nature's processes, you can convert it to CO_2 and release it back into the air. All emitted methane eventually becomes carbon dioxide anyway, so methane removal simply speeds up what happens naturally. As discussed in the Introduction, because methane is so much more potent than carbon dioxide, the CH_4-to-CO_2 tradeoff is a good one for the climate. Another advantage is that we need to remove much smaller amounts of methane than carbon dioxide to make a big difference for climate, "only" tens or hundreds of millions of tons per year for methane instead of billions of tons for carbon dioxide.

No methane removal technologies are available commercially yet, but research groups and start-ups are making progress. One approach is to blow air over a catalyst capable of converting methane to carbon dioxide. The catalyst might be embedded in a powder, a paint, or a honeycomb matrix that the air travels through. Metal catalysts, including copper and iron, embedded in polymers or specialized clays called zeolites have already shown promise for destroying methane. So have some light-activated "photocatalysts," such as zinc and titanium oxides, that are relatively cheap and widely used already in paints, cosmetics, and sunscreens. Ideally you'd design a paint to do more than "just" destroy methane; it might clean the air of other pollutants, too, such as ozone or even additional greenhouse gases such as nitrous oxide.

We don't have to start from scratch in searching for catalysts, either. Companies have spent years and billions of dollars developing catalysts to turn methane into methanol (CH_3OH), an important chemical feedstock. Instead of adding one oxygen to methane to make methanol, our goal would be to add two oxygens to make carbon dioxide (CO_2).

Researchers are also using microbes to remove methane from air. Although some microbes emit methane in wet, low-oxygen environments such as wetlands, other microbes consume it in drier soils for the energy it contains, as much as 10 percent of all methane released globally. Another approach, then, is to farm the methane-eating microbes—perhaps boosting their activity through breeding or engineering—and blowing air over them to help them consume methane at a faster rate.

Scientists are testing for microbes that consume methane quickly and grow well at low methane concentrations. Researchers in Mary Lidstrom's lab at the University of Washington recently identified a promising strain that does both things well—growing in relatively low methane concentrations and consuming methane more quickly than other strains they tested. Lidstrom's group is seeking to optimize the success of such strains, while also making sure they emit no nitrous oxide, which some methane-eating microbes do. In fact, for all methane-removal technologies we need to make sure that we don't inadvertently increase emissions of other pollutants or greenhouse gases.

You can imagine making a catalyst or other system and letting the air blow over it naturally, such as painting a roof with a methane-destroying pigment. This could be relatively cheap but also fairly slow. If you wanted to accelerate the process, you might actively push air through or across a methane-destroying catalyst, pigment, or microbial colony to increase reaction rates. Ideally, you would do that first near emission sources, where concentrations are relatively high. But blowing air is costly in terms of energy, and the energy needs to be carbon-free.

If a company is going to pay for energy to blow air across a system, combining carbon dioxide and methane removal in the same facilities makes sense to me, using blowers and air-handling systems to remove multiple gases rather than tackling one greenhouse gas at a time. We should be seeking to remove all three major greenhouse gases wherever possible—carbon dioxide, methane, and nitrous oxide—and looking for opportunities to reduce other pollutants.

Beyond engineered systems, though, there's another path to methane removal. Instead of moving the air—blowing air across a catalyst or through a microbial system—you could copy nature by releasing the catalyst and letting it move in the open air. In the atmosphere, so-called radicals of hydroxide (OH) and chlorine (Cl) destroy 90 percent or more of all methane released. If you could create and release extra Cl or OH radicals from a tower, plane, or the smokestack of a ship, you could in principle destroy methane fairly cheaply.

Most of the methane in the atmosphere is destroyed—oxidized, more accurately—by OH. This hydroxyl radical is sometimes called nature's detergent because of its role in cleansing global pollution in our air. However, most OH lasts seconds or less after forming in sunlight. Thus, even if you could generate extra OH and release it into the air, it wouldn't last long enough to travel very far. In an enclosed industrial system, though, you might be able to generate OH continuously to destroy methane in a parcel of air.

The chlorine radical (Cl) is less abundant in the atmosphere but more powerful than OH because it starts a chain reaction that destroys multiple methane molecules. To use it we would need to release more chlorine into the atmosphere in a way that doesn't create other risks, including damaging the Earth's ozone shield high in the stratosphere (see Chapter 7). Beyond potentially riskier strategies such as releasing chorine gas (Cl_2), the best way might be to mimic what nature does near the Earth's surface. Iron-rich dust that blows from drylands can travel thousands of miles over the open ocean. Iron in the dust reacts with chlorine-rich sea spray to form iron chloride aerosols that split methane in sunlight. Iron-rich dust blowing from the Sahara Desert, for instance, has been shown to increase chlorine production over the Atlantic Ocean by 40 percent, destroying millions of tons of methane.

So we could—in principle—add extra iron oxide particles into the air in the smokestacks of ships already traveling over the oceans or release iron from balloons or even aircraft. While such approaches may be technologically feasible, I am wary of tampering with the chemistry of the Earth's air at such large scales, in particular the idea of "fighting pollution with pollution." However, those of us in richer nations have already muddied the atmosphere, loading it with dangerous greenhouse gas pollution for decades. Tampering further may become necessary until such time as we can slash methane emissions, which should always be a higher priority.

Obtaining permission—and building a consensus—to release iron into the open air to destroy methane with chlorine radicals may be difficult. People are rightly skeptical about such interventions. For instance, companies would need to be sure they don't produce the wrong gases, such as turning methane into carbon monoxide (CO) instead of carbon dioxide (CO_2), forming nitrous oxide (N_2O) by mistake, or creating methyl chloride (CH_3Cl), when adding chlorine into the air. And adding chlorine directly as a gas (Cl_2) instead of adding iron to generate Cl radicals has its own risks: chlorine gas is poisonous and highly reactive.

There's a lot to get right. With all approaches, we would need to consider any health effects for people or ecosystems downwind. To understand what gases we were affecting, we would also need to measure the concentrations of methane and other gases upwind and downwind of the release point. Such measurements are more easily done in closed systems, where we can measure the concentrations of gases entering and leaving a chamber, than they are in the open air.

I believe methane removal will ultimately be needed for us to reach a safe climate, just as I now believe we need carbon removal technologies to supplement emissions reductions. As important as I think methane removal is, however, it needs much more research to examine its benefits and risks and to decide when, where, and *whether* commercial applications are warranted. We need to ensure we don't cause additional harm through technological fixes. Unfortunately, the risks are greater for the open-air manipulations that would likely be the cheapest interventions.

Like carbon removal, methane removal smacks of desperation. We're examining it solely because we've put off climate action too long.

Part III
END TO END

13

Implausible Deniability

What would you say if you believed that—through selfishness—we're in danger of ending civilization on Earth? Would you inch toward polemics and risk alienating the people you most need to reach? If you could say one last thing, what would it be?

In the Introduction, I wrote that my dream was to restore the atmosphere for methane, to harness the joy of partial victory on the path to climate survival. Controlling methane provides our best—and perhaps only—opportunity to shave peak global temperatures over the next few decades. Essentially all of the three billion tons of extra methane in the atmosphere today were released over the last twenty years. Quash those emissions, and the atmosphere will return to normal (for methane) in a decade or two.

Burning oil, natural gas, and coal emits by far the most global carbon dioxide pollution. Fossil fuel extraction and use also release one-third of all methane emissions from human activities. That combination means nothing we do is more important for a safe climate than cutting fossil fuel use. Every time we save energy by walking, biking, or combining an electric appliance with clean electricity, we cut fossil carbon pollution *and* keep methane out of our air.

Every new gasoline-powered car, every gas furnace built for our homes, every coal-fired power plant has an expected lifetime of a decade or more—with the accompanying fossil carbon pollution and methane leakage over that lifetime. If the thousands of fossil power plants on Earth today reach the end of their lifetimes without carbon capture and storage, their "committed emissions" will push us well past 1.5°C and possibly 2°C temperature increases.

With that in mind, the world cannot afford new fossil infrastructure— power plants, ships, and more. We also can't afford new oil and gas fields or coal mines. Companies know—and have long known—that oil, gas, and coal reserves will push the world well past 1.5°C and 2°C temperature increases if they're all burned. In fact, more than half of known oil, gas, and coal reserves need to go unmined for the world to stay below a 1.5°C carbon budget. Whose coal, oil, and gas will stay unburned is a question for ethicists and economists to work out. We know, though, that each company and country will lobby for *its* oil and gas to be used.

And yet, fossil companies keep digging, drilling, and blasting— ignoring the reality of climate change and the risks to shareholders of investing in unburnable fossil fuels. If fossil companies took climate change seriously, they wouldn't still be seeking more fossil fuels. Yet they are, hoping (I suspect) that they'll be able to sell what they find before someone makes them stop, and trying to lengthen that window through climate change denial and delaying tactics.

Corporate climate change denial wasn't always a reality. Ben Franta, when he was a graduate student at Stanford, dug into the history of early climate change communications within the fossil fuel industry. He showed how Frank Ikard, then president of the American Petroleum Institute (API), called industry leaders to action at the API's annual meeting in 1965. Ikard was responding to a newly released report from American president Lyndon Johnson's Science Advisory Council, titled *Restoring the Quality of Our Environment*, that highlighted the dangers of climate change from continuing fossil fuel use. Ikard never once disputed the report's findings.

That's because, beginning in 1955, the API had funded research—"Project 53"—at Caltech and elsewhere that documented the buildup of fossil carbon dioxide in the atmosphere. In his speech, and referring to the Advisory Council document, Ikard said, "One of the most important predictions of the report is that carbon dioxide is being added to the Earth's atmosphere by the burning of coal, oil, and natural gas at such a rate that by the year 2000 the heat balance will be so modified as possibly to cause marked changes in climate beyond local or even national efforts." He concluded, "The substance of the report is that there is still time to save the world's peoples from the catastrophic consequence of pollution, but time is running out." Ikard said this in 1965.

Harvard historian Naomi Oreskes has spent years documenting the gulf between Exxon's, and later ExxonMobil's, public statements from corporate executives and what Exxon scientists were telling those executives internally about climate change. She showed that Exxon scientists made strikingly accurate projections in the 1970s for how much global warming would occur with future fossil fuel burning, in many cases better than projections from professors and government scientists at the time. As Oreskes said, "It proves . . . that ExxonMobil scientists knew about this problem to a shockingly fine degree as far back as the 1980s, but company spokesmen denied, challenged, and obscured this science, starting in the late 1980s/early 1990s."

After James Hansen's landmark climate testimony before the U.S. Senate in 1988, Exxon's head of corporate research, Frank Sprow, drafted a memo for colleagues. He wrote, "If a worldwide consensus emerges that action is needed to mitigate against Greenhouse gas effects, substantial negative impacts on Exxon could occur. Any additional R&D efforts within Corporate Research on Greenhouse should have two primary purposes: 1. Protect the value of our resources (oil, gas, coal). 2. Preserve Exxon's business options." As documented in a *Wall Street Journal* exclusive on decades of Exxon strategy to "Downplay Climate Change," Sprow acknowledged that his memo was adopted by Exxon as policy.

ExxonMobil recently doubled down to protect the value of their oil, gas, and coal, as Sprow suggested, spending $60 billion to purchase Pioneer Natural Resources. The purchase more than doubled the company's footprint for fracked gas and oil in the Permian Basin. The Permian Basin yields one-fifth of all gas produced in the United States.

The deny-and-delay playbook continues. Keith McCoy, senior director for federal relations at ExxonMobil, in essence, acknowledged Exxon's efforts to slow climate action in a recent 2021 video: "Did we aggressively fight against some of the science? Yes." McCoy also acknowledged the proxy fights ExxonMobil has supported to delay climate action. "Did we join some of these shadow groups to work against some of the early efforts? Yes, that's true. But there's nothing illegal about that. We were looking out for our investments. We were looking out for our shareholders."

McCoy even suggested that Exxon's "support" for a price on carbon pollution was disingenuous: "Nobody is going to propose a tax on all Americans," he said. "The cynical side of me says we kind of know that. But it gives us a talking point. We can say, 'What is ExxonMobil for? We're for a carbon tax.'" He swiftly stated his view that a carbon tax "isn't going to happen" today.

Other companies deployed the deny-and-delay playbook to disastrous ends—for themselves and for consumers. Former senior vice president and president of Manville's Fiberglass Group Bill Sells reflected on how Manville went astray in the asbestos business: "As a manager with Johns-Manville and its successor, the Manville Corporation, for more than 30 years, I witnessed one of the most colossal corporate blunders of the twentieth century. This blunder was not the manufacture and sale of a dangerous product. Hundreds of companies make products more dangerous than asbestos—deadly chemicals, explosives, poisons—and the companies and their employees thrive. . . . In my opinion, the blunder that cost thousands of lives and destroyed an industry was a management blunder, and the blunder was denial."

What Manville should have done, Sells believes, was to take full responsibility for the manufacturing and use of their products. "Product stewardship—defined as product responsibility extending through the entire stream of commerce, from raw material extraction to the ultimate disposal of a used-up or worn-out product—can cost a lot of money. But so can the alternative," he said.

"Had the company responded to the dangers of asbestosis and lung cancer with extensive medical research, assiduous communication, insistent warnings, and a rigorous dust-reduction program, it could have saved lives and would *probably* have saved the stockholders, the industry, and, for that matter, the product," Sells added. "But Manville and the rest of the asbestos industry did almost nothing of significance—some medical studies but no follow-through, safety bulletins and dust-abatement policies but no enforcement, acknowledgment of hazards but no direct warnings to downstream customers—and their collective inaction was ruinous."

The fossil-fuel industry perpetuates Manville's mistakes today—merging decades of climate change denial with tactics to delay climate action. We all suffer the consequences.

A recent article in the journal *Global Sustainability* titled "Discourses of Climate Delay" outlined a typology of attempts to "redirect responsibility, push non-transformative solutions, emphasize the downsides of climate policy, or 'surrender' to climate change." Examples include *whataboutism*—the argument that other countries or states produce more greenhouse gas emissions and thus bear a greater responsibility for taking action, and the "free rider" excuse, which claims that others will take advantage of those who lead on climate change mitigation.

Another common delaying tactic is *individualism*, which redirects climate action from systemic solutions to strictly individual actions, such as driving a more efficient car. As the authors describe it, "This discourse narrows the solution space to personal consumption choices, obscuring the role of powerful actors and organizations in shaping those choices and driving fossil fuel emissions."

Reducing our individual carbon footprints helps but won't solve the climate problem without systemic changes in energy supply, heavy industries such as steel, and consumption. When companies highlight the importance of individual actions to reduce our emissions, they shift blame from their own responsibilities.

I believed British Petroleum when they launched their "Beyond Petroleum" campaign more than two decades ago (and—as cynics pointed out—right after their $48 billion acquisition of oil giant Amoco, the largest industrial merger to date). Advertising agency Ogilvy & Mather developed the "Beyond Petroleum" campaign for hundreds of millions of dollars and eventually recommended renaming the company "beyond petroleum." CEO John Browne announced the change at a speech at Stanford in 2002, saying "We need to reinvent the energy business; to go beyond petroleum."

As part of the advertising campaign, Ogilvy & Mather created the "carbon footprint calculator" for BP, with the admonition "It's time to go on a low-carbon diet." The website took readers to a site where they could upload their recent food, shopping, and travel history. The site estimated the reader's greenhouse gas emissions and individual planetary impact.

In one archived television ad from the fall of 2003, Ogilvy and Mather interviewed people on the street: "What size is your carbon footprint?" read the screen in simple black letters to the right of the new-at-the-time green-and-gold BP logo.

The ad switched to four people answering the question:

"Ahhh, the carbon footprint. That I don't know."

"Whatever it is, the whole population of the world make that a very, very big number."

"How much carbon I produce. Is that it?"

"You mean the effect that my living has on the Earth in terms of the products I consume?"

Then the ad switched back to a white screen with stylish black fonts:

"We can all do more to emit less." Then:

"Over the next 4 years, we're planning to implement projects to reduce emissions by another 4 million tons."

"Learn how to lower your carbon footprint at bp.com/carbon footprint."

"It's a start."

The ads were genius. A year after the first ad launched, nearly three hundred thousand people entered their data on BP's website to estimate their own carbon footprints.

In early 2006, BP's updated website once again read, "It's time to go on a low-carbon diet." Advertising executive John Kenney wrote an opinion piece in the *New York Times* that year titled "Beyond Propaganda." Kenney described how "six years ago I helped create BP's current advertising campaign, the man-in-the-street television commercials." He said: "Advertising is a funny business. You get to help shape the personalities of huge companies. Most often it's for cellphone service or credit cards or fast food or paper towels. Rarely are you faced with whether you 'believe' in a product or service. This was different. This was serious. I believed wholeheartedly in BP's message, that we could go—or at least work toward going—beyond petroleum."

He mused differently later in the op-ed, however: "I guess, looking at it now, 'beyond petroleum' is just advertising. It's become mere marketing—perhaps it always was—instead of a genuine attempt to engage the public in the debate or a corporate rallying cry to change the paradigm." Kenney continued to the end: "Maybe I'm naïve. Think of it. Going beyond petroleum. The best and brightest, at a company that can provide practically unlimited resources, trying to find newer, smarter, cleaner ways of powering the world. Only they didn't go beyond petroleum. They are petroleum. The problem there is that 'are petroleum' just isn't a great tagline."

In October of 2018, BP spent more than $10 billion to purchase petroleum assets in the fracking-rich Permian and Eagle Ford basins. The acquisition added hundreds of thousands of barrels' worth of oil and gas production a day to BP's portfolio and almost five billion oil-equivalent barrels of "recoverable" petroleum in total.

The BP press release dropped any pretense of the Beyond Petroleum campaign and replaced the campaign's simple language with almost Orwellian corporate-speak: "Following integration, the transaction will be accretive to earnings, is estimated to generate more than $350 million of annual pre-tax synergies and is expected to boost Upstream pre-tax free cash flow by $1 billion, to $14–15 billion in 2021." Bernard Looney, chief executive of BP's Upstream business, said, "We look forward now to safely integrating these great assets into our business and are excited about the potential they have for delivering growth well into the next decade."

Beyond Petroleum indeed.

Continued corporate temporizing on greenhouse gas cuts will be ruinous. In addition to new policies, regulations, and incentives to fight climate change, we need success stories, examples of companies and people who are genuinely changing their behaviors.

Ørsted, once known as Danish Oil and Natural Gas (DONG) Energy, has transformed itself from a conventional fossil power company to a global renewables powerhouse. In 2008, it produced power in the ratio of 85:15 fossils to renewables. According to Martin Neubert, CEO of Ørsted's offshore wind business, the impetus for the company's change came from public resistance to a new coal plant in northeast Germany. Ørsted failed to win local approval after spending six years developing the project and receiving full support from the German government. "This was the first clear sign telling us that the world was beginning to move in a different direction," he explained. Neubert also acknowledged the importance of the U.N. Climate Conference held soon after in 2009 in Copenhagen, a conference that highlighted the renewable-energy transition.

DONG decided to change from a conventional fossil-based energy company to a renewables business "within one generation." Remarkably, they flipped their ratio of 85:15 fossils to renewables in a single decade instead of their initial target of more than three decades. They did it by becoming the global leader in offshore wind energy.

Neubert described the transition this way: "The first step was divesting our oil and gas business, which concentrated our business almost entirely on renewables." According to Neubert, Ørsted plans to end its coal use by 2024 and have its power generation be carbon neutral by 2025.

The transition wasn't easy, though, and the company faced resistance from employees. As Neubert described it, "There was internal pressure to keep DONG Energy the same. It wasn't unexpected, because we had spent three decades turning the company into a traditional fossil fuel company. Fossil fuels were our core competence and the focus of our growth strategy. . . . The skepticism was broad and profound. Ultimately, though, internal skepticism receded."

DONG went public successfully in 2016 and renamed itself Ørsted to reflect the lack of fossil assets in its portfolio and to honor Danish inventor Hans Christian Ørsted, who discovered the electromagnetism that drives wind power today. Ørsted Energy now builds onshore wind, solar PV, and electricity storage along with its offshore wind assets. Today, Neubert says, "everything we do reflects our vision to create a world that runs entirely on green energy."

Creating a world that runs "entirely on green energy" requires systemic changes in energy supply. That transformation can be driven by reshaping business practices at companies such as Ørsted and—maybe someday—ExxonMobil, BP, and others.

14

Some Heroes to Zero

I f companies can change in a decade, people can, too. And though reducing our personal emissions helps, all of us who care about maintaining a livable climate need to scale our actions well beyond our
own lives and homes. That scaling could happen in organizing your
block or neighborhood—your church, school, or business. All of us can
run for city council, organize voter initiatives, or start a climate-friendly
group to scale climate solutions beyond our individual consumption. I
highlight three examples of special people doing their share.

Mats Karström is a high school teacher who has spent nearly forty
years protecting Sweden's old-growth forests and the rare species
found in them. I visited Karström at his home in Jokkmokk—Swedish
Lapland—to discuss his work and to learn about Steget Före (or "One
Step Ahead"), the forest conservation group he founded decades ago.
Through Steget Före, Karström has trained thousands of citizen-science
volunteers to inventory forest plots for their rarest species.

I pull into Mat Karström's driveway in the town of Vuollerim with
friends and ecologists Anders Ahlström and Julika Wolf; we're scouting
old-growth forests for future research on carbon storage and biodiversity. When we arrive, Karström is standing in his garden gripping a butterfly net in a two-fisted tennis backhand. The month before he found a
butterfly in his yard that had been seen only once before in Sweden.

Karström invites us into his three-hundred-year-old farmhouse covered in red vertical siding with white-trim posts and gingerbread porch railings. His home resembles a small-town museum with images of species—rare and not so rare—everywhere: paintings of calypso orchids; photos of the *vitryggig hackspett*, or the white-backed woodpecker, an indicator species Karström found that requires large forest tracts to breed; the mushroom field guide (*Svampar i Sverige* or *Swedish Mushrooms*) he coauthored, with its yellow-chanterelle cover; a painting of two new species of wood midges with scientific names that honor Karström and the forest organization he founded: *Peromyia karstroemi* and *Campylomyza stegetfore*.

After moving to Lapland in the 1980s to teach high school, Karström grew concerned about unchecked logging nearby and realized there were no comprehensive maps of rare species for the region. "The thing is, if there isn't anything of conservation value to save, then the forest is considered worthless, and they don't have to take care of it," he says. "The logging companies and Swedish conservation agencies didn't even know which rare species were found here."

After learning the rare species with help from local naturalists, Karström surveyed dozens of forests himself, often with his students. "When I first came here," he says, "I went into the field maybe two hundred and fifty days a year."

From his initial surveys, Karström developed a Pyramid of Protection that rates forests for their conservation value using the species and attributes of the trees and the presence of rare species among them, including birds, herbs, flowers, lichens, and fungi. His pyramid has been widely adopted by Swedish forestry companies and the Swedish Environmental Protection Agency as a way to rate old-growth forests. The base of his pyramid holds younger, less diverse forests with the fewest rare species. The top of the pyramid marks old-growth forests with the greatest number of rare species, forests he playfully calls "Nirvana."

Karström taught his biology students to identify rare species. "I began with birds because everyone knows a few birds, and I didn't want to scare my students away. Later, I taught them rare flowers and plant species—then lichens, mushrooms, and mosses." He rewarded his students with ice cream when they learned to spell complicated scientific names.

"The normal process is for students to learn the names of the common species," Karström says. "But I realized you don't have to be a biologist to identify the fairy slipper orchid, *Calypso bulbosa*, and other rare species." Over the next decades he trained six thousand students, parents, and more to identify the rarest species of concern rather than the most common species. He said, "If we found eight red-listed species, then I could scare the logging companies with them."

He recruited other partners, including moose hunters. "They spend time in the forests and they care about habitat protection. I had them hunting lichens instead of moose meat," Karström says with a smile.

His volunteer army scoured Lapland's forests to document rare species. "My point system was a way to excite people," he says. "I could give it to anyone, and they didn't have to be in university to use it." He and other experts checked the species-rich plots their volunteers identified and mapped the locations.

The public maps put pressure on forestry companies in the area and on the Swedish government to save what old-growth forests remained. "When we started in 1986, we had the worst forestry in Sweden with the biggest clear-cuts, and no one knew anything about local conservation values," Karström says. "Today, it's different. With our maps, people know when a forestry company has cut an old-growth forest. 'You have been bad,' I can say through journalists. Then I submit five more forests for protection."

"So far, we have protected more than one hundred and fifty forests," he says. "They are my children."

Like children, his forests come in different sizes. Last summer, Karström and his partners finally succeeded in preserving the Jielkká-Rijmagåbbå forest and wetland area to the north. He beams. "I worked thirty-five years to save it," he says. "My biggest forest—five hundred square kilometers—was privately owned but now it's government owned. I haven't completely grasped it yet. It's nearly as big as Muddus National Park nearby." The new preserve protects eighteen rare habitat types and many rare species.

"I want to tell this story," Karström says, "for people who come after me. They need to see that local action more than the government saved these forests. It's important for young people to meet the companies. They should play a part in forest protection."

Mats Karström has good relationships today with both the forestry companies and community members. "Today I have a good relationship with SCA [Svenska Cellulosa Aktiebolaget], the largest local forestry company. In the beginning they were threatening. Today they are nice. They understand it's a better way."

Scientist Anders Ahlström was more negative about Swedish forestry policy. "We're still cutting our oldest, most beautiful forests and turning them into toilet paper," he says. "It makes no sense to cut the last remaining 8 percent of old-growth forests." That's why he appreciates Karström's efforts so much—they combine conservation *and* climate solutions.

After sharing home-cooked ginger snap cookies with us, Mats Karström *finally* asks if we want to visit one of the protected forests. Arriving at a spruce and pine forest only half an hour from his home, he bounds out of the car and opens his arms in welcome. "There are calypso orchids here," he says, "but they aren't flowering now." He bends over and picks a small white-green lichen. "*Ramalina thrausta!*" he says—in truth sounding more like "Rom-a-Leee-na tha-rowww-sta!"—"It's rare here." He shows us dozens of species. He points out a red-listed "Lactarius!" mushroom. "Overall, I've found thirty red-listed species in this forest," he says. Karström's enthusiasm infects geoscientist Julika Wolf, who springs off to collect an edible Porcini mushroom she spies.

But then Karström finds a rutted spruce tree, dead, with little paths bored randomly across its bark. "Oh, no," he says. "Bark beetles." He notes how unusually dry the forest floor, trees, and lichen are. Brown spruce cones litter the ground. "Something's wrong," he says. "Lots of cones turned brown this year and dropped to the ground. There's stress everywhere. I've been here thirty-six years and never seen this or the beetles before."

Like so many people I interviewed for this book, Mats Karström sees unprecedented changes happening *now*. "I never used to think that climate change was a big problem for my forests. Maybe in fifty years. But it has happened in places this year for the first time. We have such huge temperature swings. It's also getting very, very hot at the start of summer."

Online, I later found scientific papers and news headlines confirming Karström's view, particularly for Sweden's destructive bark beetle boom that accelerated around 2019, as warming temperatures allowed the insects to expand into new habitats. One 2019 story from the University of Würzburg reads: "Bark beetles are currently responsible for killing an unprecedented number of trees in forests across Europe and North America." Another, from 2020 was titled "The Bark-Beetle Invades Europe."

Karström notes the irony. "After 'protecting' so many forests in my lifetime, I could live to see them lost to climate change anyway."

Meet Reverend Lennox Yearwood, Jr., president and CEO of the Hip Hop Caucus. "Rev," as he prefers, sits before me wearing thick, dark-rimmed glasses and a black hat that reads "END FOSSIL FUELS NOW" in white block letters. Yearwood wears many hats, including: "Respect My Vote!," "#ExxonKnew," "#NOKXL" referring to the Keystone XL pipeline, and "NY" in stylized Yankees script, a hat most unfortunate to fans of other baseball teams. Yearwood and I serve together on the Board of Advisors to a climate nonprofit.

Yearwood's not your typical climate organizer. While teaching social justice at Georgetown University two decades ago, he was simultaneously protesting the Iraq War, which, he acknowledged, "wasn't the best career move for someone in the air force."

"I went to a lot of rallies," he said, and saw different communities relating to various issues. "If I went to the police brutality rally, it was all Black, the immigration rally was all brown, the anti-war rally—all white. And I'm like, 'Dang, this ain't gonna work because we're all one people trying to challenge the same entities and industries.'"

He singled out the climate movement, in particular. "What I noticed is that it's a predominantly white East Coast–West Coast kind of Birkenstock crowd, to be honest." (Having never worn Birkenstocks, I still take his point.) And this comment comes from someone who counts climate champion and northeasterner Bill McKibben as a "hockey buddy" and "one of my best friends in the world." (I don't know whether McKibben wears Birkenstocks. . . .)

"The climate movement needs to broaden itself," Yearwood said. "I figured we needed to change that before we could really change anything else. In hip-hop, you have black and white and brown and red people connecting through musical culture. We wanted to use music to reach rich and poor people, rural and urban people."

Yearwood helped form the nonprofit, nonpartisan Hip Hop Caucus in 2004 "to fight poverty and pollution at the same time." He also sought to strengthen political activism for young voters using hip-hop music to reach them. He said, "Caucus members understood the importance of culture in building audiences and bringing people together. Our bedrock idea was 'How can we use cultural expression to shape political experience?' Could we use it to create real change around difficult issues, from voter suppression to climate justice?"

Beyond climate change, voting has been one of the Caucus's longest-running priorities, such as its successful "Respect My Vote!" campaign. The Hip Hop Caucus has brought tens of thousands of young voters to the polls with the help of artists such as Common, Dawn Richard,

Ludacris, Omar Epps, and Jay-Z. In 2008 Yearwood and the Caucus registered thirty-two thousand voters in sixteen cities, setting a world record for voter registration in one day. *Rolling Stone* dubbed him a "New Green Hero." The Obama White House called him a "Champion of Change."

The Hip Hop Caucus also won early awards for its hurricane recovery efforts after Katrina drowned New Orleans in 2005. Born in Louisiana, Yearwood said, "Hurricane Katrina hit the Gulf Coast after having already gone through Florida. And we know now that warming made it worse. It went from literally a Category 1 to a Category 4 and 5 storm very quickly." Scientific papers support Yearwood's assertion that climate change strengthened Katrina, including decades of sea level rise that worsened the flooding by an estimated 15 to 60 percent. Unusually warm ocean surface temperatures also supercharged Katrina's rainfall and intensity.

Yearwood lamented the "people destruction" of Katrina the most, "seeing communities left behind in the richest country in the world. . . . I was in Washington, D.C., when Katrina hit. Seeing family and friends drown, and then watching the survivors get left behind, made me realize, 'We have to do a better job of dealing with the climate crisis.' It was heartbreaking."

The Hip Hop Caucus helped displaced people from New Orleans return home. It also promoted environmental justice in their paths to recovery.

Three years later, Ida struck New Orleans as one of the strongest hurricanes ever to hit Louisiana. Yearwood and his Caucus created the Hurricane Ida Relief Fund that awarded snap grants to people by Venmo. "People were literally on the run," he said. "If they signed up and said they needed $100, they got it instantly. People needed money for diapers and hotel rooms. Because of the climate crisis, we are having to rethink how we work with humanity and make sure people are safe."

Yearwood sees the climate crisis far beyond harm from hurricanes:

"Climate has put many, many, many communities on the front lines. We're seeing this in the wildfires hitting the West Coast, the hurricanes ravaging the Gulf South, the floods in Pakistan and India, the recurring droughts and heat waves parching Australia and the United States. The climate crisis impacts every part of the world."

The harm isn't equally distributed, though. Echoing Catherine Coleman Flowers in Chapter 1, Yearwood said, "Climate change's biggest impact is on poor communities, particularly black, brown, indigenous, and poor white communities. The poor are in harm's way."

In addition to his experience as a community organizer, Yearwood is the senior advisor to Bloomberg Philanthropies, Michael Bloomberg's climate-active foundation. Yearwood chairs Bloomberg's "Beyond Petrochemical" campaign. "At Bloomberg," he said, "we promote electrification—from homes and buildings to making sure there are enough EV chargers in rich and poor neighborhoods, both. We apply an equity lens, too, seeking an equal distribution for climate solutions across zip and building codes. We want to make sure that all communities benefit from grants to put solar panels on homes, build new bike lanes, and improve their communities. We're also investing in ways to reduce methane and carbon dioxide emissions around coal, gas, and petrochemical facilities."

Yearwood believes in the coalition the Hip Hop Caucus is building. "I work with young people all the time at the Caucus," he said. "They have a 'We can't lose' mentality because really, they can't. They realize that while many of their parents fought for equality in the twentieth century, they're fighting for survival in the twenty-first century, and losing isn't an option."

Yearwood also supports public demonstrations and new champions of climate justice who are emerging. "Think of Catherine Coleman Flowers," he said, "and the way she's dealing with water and sanitation issues in Alabama. It's so important for communities to come together and say, 'This is not acceptable.' Think of Sharon Lavigne in St. James Parish and Cancer Alley fighting the Formosa plastic plant and saying, 'No!'"

"I'm not wearing rose-colored glasses," he added. "Fossil energy is probably the hardest thing people have ever had to transition from. I understand the power of the fossil fuel industry. I'm aware that their business plan is a death sentence for our communities.

"I still believe we're going to succeed. I've seen where we can get to. It's important to know that when your ancestors came to this country as slaves, and now you have folks who are president, vice president, it gives you hope against tremendous odds."

"And I don't think billionaires will solve this problem, a distant carbon capture kind of thing far down the road. People coming together is the solution. Wherever the people are together, I will put my chips on that right there."

Rose Abramoff is a Millennial soil ecologist and rising star scientist who studies forests, peatlands, and tundra systems across the globe. She has led groundbreaking research testing the effects of global warming on plants and soils. She and I have worked together to measure soil organic carbon and boost long-lived soil carbon pools.

I met Abramoff most recently while attending a meeting of the American Geophysical Union. We met in a coffee shop as Låpsley's *Falling Short* played overhead. To avoid greenhouse gas emissions from flying, she had just ridden the train cross country from Maine to San Francisco. "I treated myself to a Roomette on the Amtrak," she said, almost apologetically.

Like many of us, Abramoff credits personal experiences with galvanizing her call to climate action. Her hands turned actively like a windmill when she spoke. "Many times doing field work were harrowing," she said. "Once, I was collecting data from a warming experiment on the North Slope of Alaska in Utqiagvik [formerly "Barrow"]. I was doing radiocarbon and isotope work while measuring carbon dioxide and methane emissions from experimental plots.

"The summer I went in 2016 was the year Alaska shattered its record for warmest year. One of the things we track each summer is how deep the active layer is—the maximum depth that Arctic soils thaw in summer before refreezing in winter. Usually in Utqiagvik, the active layer is only about six inches deep in summer. But this time, I pushed the rebar into the soil, and it kept sinking lower and lower. That was just crazy. Meanwhile, I saw slumps and subsidence on the lands all around me. You felt impending disaster."

These experiences galvanized Abramoff to action. "I've always been a politically conscious person," she said. "I participated in a lot of soft activism in my life—you know, street protests such as the March for Science, Occupy Wall Street, that sort of thing. But I'd been reluctant to do riskier activism, mostly because I was trained to pretty much stay in my lane, to follow the rules and 'do what I'm supposed to do.' I thought, 'Okay, I'll write papers and contribute to reports and do educational activities,' all the usual stuff professors do.

"Then I went to France, which has more of a protest culture. I was still reluctant to do too much there, though, because I didn't have rights of citizenship. But I started volunteering for Scientists for Extinction Rebellion."

The ticking time bomb of climate data pushed her further and faster. Rose Abramoff is one of the climate experts who reviews the latest climate assessments for the IPCC, the United Nations body that investigates climate change. "Going through all of that data—so many hard numbers—I was like, 'Shit, we don't have time.' I thought, 'As soon as I get home and regain the rights of citizenship, I will be more active.'"

"We're fighting for our lives," she said. "Originally, I thought, 'Okay, I'll wait until I'm James Hansen—a distinguished professor of whatever the hell—and then I'll chain myself to the White House just like he did, and everyone will listen to me,' which is of course not what happened with him. But every time I read a new report, I sped up my timeline for action. We have no time to wait."

James Hansen was Abramoff's inspiration on April 6, 2022, when she and four activists from Declare Emergency and other groups chained themselves to a White House gate attempting to get President Biden to declare a climate emergency. They weren't alone. More than a thousand members of Scientist Rebellion in dozens of countries risked arrest that day to protest inaction on climate change. Here was the group's statement in an open letter:

> We are scientists and academics who believe we should expose the reality and severity of the climate and ecological emergency by engaging in non-violent civil disobedience. Unless those best placed to understand behave as if this is an emergency, we cannot expect the public to do so. Some believe that appearing "alarmist" is detrimental—but we are terrified by what we see, and believe it is both vital and right to express our fears openly. . . . The most effective means of achieving systemic change in modern history is through non-violent civil resistance.

Rose expected to be arrested at the White House on the day of "Declare Climate Emergency." She felt the personal inconvenience warranted the attention and conversations the groups' actions fostered. She noted—without directly crediting their protest—that soon afterward President Biden authorized the U.S. Defense Production Act to speed the transition to renewables. As described in a Department of Energy announcement, Biden provided the DOE with the authority "to accelerate domestic production of five key energy technologies: (1) solar; (2) transformers and electric grid components; (3) heat pumps; (4) insulation; and (5) electrolyzers, fuel cells, and platinum group metals."

"I chose civil disobedience," Abramoff said, "not because I think it's the most desirable path forward but because of the powerful symbolism of scientists showing there's a climate emergency. I also look to non-violent historical movements from the past, which include the Indian Democracy Movement, Civil Rights, the Suffragettes, and South African apartheid. Their success gives me confidence."

"What would you like scientists like me to do?" I asked, with some trepidation.

"Exactly what I'm doing now," she said, referring to protesting and other direct actions. When I demurred, she added "Give me a call when you're ready."

She continued. "I'd like to see scientists shift how they see themselves and their policy neutrality—seeing themselves more as part of their communities. It's the idea of shedding policy neutrality and stepping down from the ivory tower. In joining the activist community, I've had a real sense of joining the rest of humanity. It feels like we are finally and belatedly joining an existing movement that has been led for decades by youth and Indigenous people.

"Find people in your community," Abramoff said, in a call to action. "Find a small group of people who align with your goals and meet with them. And if you need more guidance you can tap into a more established group. I tapped into Scientist Rebellion, but now I work with groups that organize in other ways. Don't chain yourself to a gate alone. It's important to create a support structure before you do anything."

Abramoff calls for institutional change and highlights contradictions she sees in what's permitted. She worked as an ecologist for Oak Ridge National Laboratory, an energy lab maintained by the U.S. Department of Energy. "Oak Ridge and other national labs all employ lobbyists. But the only thing they lobby for is more research funding. That's fine—research funding is important. But there's no rule that says we can't lobby the government for maintaining the habitability of the Earth or for things that might facilitate a better, more equitable future."

Abramoff ramped up her activism in the fall of 2022. She and three others were arrested protesting greenhouse gas pollution at the private jet terminal of Charlotte Douglas International Airport in North Carolina. Peter Kalmus, a scientist at the Jet Propulsion Laboratory who was arrested with Abramoff, said, "I feel desperate. I feel like doing this for the planet, and for my kids and for young people is, frankly, more important than being safe in my career. So, I do everything I can not to jeopardize my job. But sometimes there's, you know, basically a higher calling."

A month later, Rose Abramoff and Peter Kalmus walked on stage at a scientific session of the American Geophysical Union held at Chicago's McCormick Place, the largest convention center in North America. While hundreds of scientists waited in the ballroom for the program to begin, Abramoff and Kalmus unfurled a banner that read "OUT OF THE LAB & INTO THE STREETS." They each read brief statements. Kalmus spoke first: "Our science is showing that the planet is dying—it's terrifying. Everything is at risk. As scientists, we have tremendous leverage, but we need to use it. We can wake everybody up." Abramoff spoke one sentence—"Please, please, please find a way to take action!"—before security and AGU staff escorted them off the stage.

She described the fallout: "They kicked us out of the conference, and I received a phone call from one of the organizers after we left telling us that if we returned to the conference center, we'd be arrested. They removed our scientific research from the program, too—my two posters and Peter's talk. Oak Ridge National Lab also fired me," she added.

Yes, Rose Abramoff *had* a permanent position at Oak Ridge National Laboratory. Describing the firing, she said, "To be fair, I wasn't totally shocked. I knew even before April that my actions, especially the nonviolent civil disobedience, risked my job."

I wasn't shocked hearing that Oak Ridge fired her, either. The national labs have strict rules about political statements for people on official business, even if they say (as Abramoff does in her X account), "Opinions expressed are my own."

She would do it again, though. "So far I haven't regretted any of my choices," she said. "Maybe if I'm sent to prison for a year, I'll feel differently." At the time we spoke, Abramoff was facing charges in West Virginia for trespassing, obstruction of justice, and interfering with property rights after chaining herself to the Mountain Valley gas pipeline under construction. She explained her rationale for acting: "I try to make decisions imagining myself as a woman at the end of her life looking back on it. What are the things I can do now that will make me proud of my life? I just thought it was important enough to take the risk."

Despair notwithstanding, she said, "I'm still optimistic, though. Do I think we're going to meet all of our climate goals? Probably not, but can our collective actions lower peak temperature to something that's more survivable? Yes, absolutely. Let's do that. Let's shave every tenth of a degree off that we can. And who knows? Maybe we'll manage to generate some overwhelming social mandate to create a World War II–scale mobilization.

"Oak Ridge Lab was built entirely between 1942 and 1944," she said, describing the lab's origin as part of the Manhattan Project. "They built the entire town to house all the people. They built the whole laboratory and put functioning atomic weapons together in less than three years. So, we should be able to complete a green energy transition this decade. There's no reason we can't do that."

"And if it means rocking the boat a bit?" I asked.

"I'm all for decorum," she said, "but not when it costs us the Earth."

15

Into the Clear

Our boat rocks a bit when a pink river dolphin surfaces and blows with a *whoosh* just a paddle's length away. We're in a low, open boat in western Brazil, a few hundred miles downriver from the borders of Colombia and Peru. I'm with colleagues at the Mamirauá Sustainable Development Reserve deep in the Amazon to build new towers where we will monitor methane and carbon dioxide emissions. The highest methane emissions in nature come from tropical wetlands and seasonally flooded forests—like here at Mamirauá—and they are expected to rise with warming. I'm also sampling in the Amazon to view climate change's reach: if the Amazon is vulnerable, then everywhere (and everyone) is vulnerable.

Mamirauá is the western jewel in a chain of national parks and reserves placed deep in the Amazon to protect biodiverse forests, and the primates, people, and fish who call them home. Mamirauá includes indigenous lands and promotes the well-being of the *ribeirinhos*—traditional peoples living in communities along the river who fish, farm, selectively cut trees, and make crafts for their livelihoods—hence its designation as a "sustainable development reserve."

Mamirauá is a sanctum of water, sky, wetlands, and trees. The water is rich and tea-stained. The air is redolent of mud.

Tropical wetlands yield so much methane because they are warm, wet (by definition), and low-oxygen environments perfect for growing methane-emitting microbes. But tropical wetlands and flooded forests are also the world's least studied ecosystems for methane emissions—and almost everything else, for that matter. They're hard to reach and a challenge to keep instruments running in, thanks to factors that range from the constant humidity to rampant ants that can short wires.

Tropical methane emissions are expected to increase because warmer temperatures boost microbial activity. Our new towers will help us measure methane emissions today and provide a baseline for documenting future changes. Recent work suggests that some wetlands may *already* be releasing more methane because of warmer temperatures. Scientists using satellite data concluded that higher tropical wetland emissions from the Amazon and elsewhere yielded most of the increases in atmospheric methane observed over the past decade. Without carbon and methane removal technologies, the world has no way to address such increases. You can turn a wrench to quench methane emissions in an oil field, but there is no wrench to turn for the Amazon.

The Amazon is in danger of unraveling from a combination of human pressures and climate change. Beyond deforestation, illegal gold mining is surging. Illegal mining destroys biodiverse forests and contaminates watersheds with the mercury used to harvest the gold. As a result, forty percent of people tested in a hundred villages in the Peruvian Amazon had dangerously high levels of mercury from eating fish contaminated with mine waste.

Human pressures are building from outside the Amazon, too, through climate change. New research shows that tropical forests are approaching critical temperature thresholds beyond which leaves will die from the heat. Such "tipping points" may kill rainforests over large areas and keep them from regrowing.

An Amazon kingfisher flashes white as it rattles across the channel in front of our boat. I turn to my host, Brazilian hydrologist Ayan Santos Fleischmann. Born in the city of Porto Alegre in southern Brazil, Fleischmann directs Mamirauá's geosciences research. He leads data collection for water, air, and climate change. I ask him whether and how climate change is altering Amazonia.

"Everyone acknowledges it's hotter and harder to work," Fleischmann says. "In the past, people would go to work in the field at 2 p.m. Now they need to wait till 3 or 4 because the heat is unbearable."

" 'The Amazon' is really many Amazons," he adds. "We currently have what you would call a hydroclimate dipole—more rainfall north and west of us next to the Andes and less rainfall and more drought in the lower Amazon to the south. This is a very clear trend in recent decades."

Fleischmann says, "We're already having more rainfall and floods in the upper Amazon." His group published a recent paper showing a one-quarter increase in the maximum flooded area across the Amazon floodplain since the 1980s—a 26 percent increase in only forty years. "That means we have thirty thousand square kilometers of more flooded area now than in the eighties. Combined with warmer temperatures, more inundation and flooding should increase methane emissions," he says.

As I wonder how the "other Amazons" are changing, particularly the lower Amazon, we float into a vast wetland that we'd marked on a satellite map. A ten-foot caiman eyes us, the slits of its pupils standing tall as trees that open into darkness. Horned screamers, or *camungos*, squawk their awkward calls from waterside branches as we pass, their unicorn horns bobbing. But the real stars are flocks of orange-mohawked hoatzin. Hoatzin are unique in having claws on their wings, like prehistoric archaeopteryx, so chicks can hold on to tree branches to keep from falling in the water. Their *chuffs* surround us.

Hoatzin are also the world's only flying cows, hosting microbes to ferment their leafy diet. Not surprisingly, they burp methane, too. And although, yes, I'm curious, I haven't traveled to the bowels of the Amazon to measure methane emissions from a bird.

Fleischmann brings me back to the risks of droughts and fires that receive global attention for Amazonian climate change. "The worst

Amazon droughts happen in El Niño years with warm Atlantic waters nearby," he explains. The key ocean region that influences Amazon droughts is between 0 and 20 degrees north—roughly the belt from the equator to Cuba and southern Florida. The extreme drought triggered by the 2015–2016 El Niño featured record high temperatures that killed billions of trees—turning the Amazon from a global carbon sponge to a vast carbon source. Amazon fires raged in both 2015 and 2016.

And now here we are in July of 2023, as El Niño strengthens again and the tropical Atlantic is baking. Ocean temperatures off the coast of Florida during my Amazon stay topped hot-tub levels of 101°F—close to temperatures suggested for cooking Atlantic salmon and the highest surface ocean temperatures measured. "This appears to be unprecedented in our records," said NOAA scientist Derek Manzello, coordinator of NOAA's Coral Reef Watch, as reefs died and health monitors warned people not to swim because of the risks of overheating and exhaustion. Legendary distance swimmer Diana Nyad—one of the first people to swim from Cuba to Florida—wrote a column titled "What It's Like to Swim in an Ocean That's 100 Degrees," reflecting on the record ocean temperatures: "Years from now, we may well remember the summer of 2023 as the beginning of an era when many of our oceans stopped serving as a glorious place of recreation."

Because of the strengthening El Niño and warm ocean temperatures while I was there, Fleischmann and I discussed how late 2023 and 2024 could both be years the Amazon—the southern Amazon, especially—fries, dries, and burns. "Water levels several hundred kilometers upriver at a monitoring station in Tabatinga, Brazil, are already as low as they've ever been." Fleischmann says from our boat, "Drought may be coming."

His warning was prescient. In late September—just two months after I left—the region baked in unprecedented drought. Water levels in the Amazon system were lower than at any time since record-keeping began in the early 1900s. Brazil's minister for the environment, Marina Silva, said, "We are seeing a collision of two phenomena, one natural, which is El Niño, and the other a phenomenon produced by humans, which is the change in the Earth's temperature." She added that this drought is "incomparably stronger" than in the past and that such droughts "could happen more frequently."

Air temperatures around Mamirauá topped 40°C (104°F) for days, and the absence of rain and clouds cooked Amazon waters in the sun. In Lake Tefé—a tributary of, and gateway to, the western Amazon where I first met Fleischmann—he measured water temperatures at an astounding 105°F at depths of three to six feet.

When we spoke over Zoom a few days later, Fleischmann also wept. "Nobody ever saw anything like this before. The first day, September 25, I saw ten river dolphin carcasses. That was a shock. Then two days later I saw 70 carcasses along the lake and one animal still agonizing." He paused to collect himself. "It was about 4 p.m. and very hot. I watched a dolphin swimming in circles—struggling to survive. It was horrible. We didn't know what to do or how to help it. If you try to rescue an animal that is already hurt it can die from the extra stress."

Not only was it hot and dry, but more than seven thousand fires raged across Amazonas state. "On that terrible day," Fleischmann said, "we had a huge fire here with high concentrations of particulate matter we measured at the institute. That pollution could be another source of stress for the dolphins as they surface to breathe air. We're also doing autopsies to test for signs of diseases."

People were suffering, too. Many *ribeirinhos* couldn't get to hospitals or find food or water because water levels were too low to allow boat travel. News photos displayed dead dolphins hauled onto tarps next to images of desperate people hand-digging wells for drinking water in dry riverbeds. Dead fish paved riverbanks in silver-brick roads. "Without water, there is no life," said a local fisherman.

I feel Fleischmann's despair—plus anger I didn't feel a decade ago. I'm sadder and increasingly discouraged about where we are in 2024.

What actions seem "reasonable" today look so different to me than when I began studying climate change. I never thought I would live to see the world's weather unravel and people suffer so much for it.

During our first conversations in July, Fleischmann had been pragmatic in his views on climate change. "As a Brazilian, I find talk of climate anxiety strange because that is not the reality here. We have many problems. People are starving. Many people don't have jobs, don't have work, don't have money for their basic needs. We lack decent sanitation in much of the country, so there are other priorities along with climate change. Nevertheless, the consequences of climate change will be devastating."

Two months later, he was shaken. "This is a tragedy in the Amazon without precedent," he said. "It may be the first sign of a tipping point. The Amazon is truly in danger."

We are all in danger. I reflect back on all the places in this book where climate records shattered during my visits: my trip to the Amazon as Atlantic surface waters boiled; the Vatican during Europe's hottest-ever summer that killed sixty thousand people overall and eighteen thousand in Italy; the sweltering heat wave in the northwestern United States and Canada that shattered all-time temperature records. Lytton, British Columbia, set Canada's highest temperature ever recorded at 121°F (nearly 50°C) and promptly burned to the ground the next day.

And as drought withers the Amazon, the month I was there turned out to be the single hottest month ever worldwide—July—in the hottest year on record—2023. Canada suffered its most destructive fire season, worsened by unusually hot and dry conditions. Forty-six million Canadian acres burned by October—5 percent of all Canadian forests and nearly triple the amount burned in any previous year. Dangerous wildfire pollution choked people from Ottawa to Miami. A longtime wildland firefighter and Canadian fire chief surveyed the country's destruction and said, "The unimaginable has to be imaginable now."

When we in the Global Carbon Project published our yearly estimate of fossil carbon emissions for 2019—a record at the time, of course—I was quoted in the *Washington Post* saying, "We're blowing through our carbon budget the way an addict blows through cash." That was five years ago and nothing has changed—not even COVID-19 could slow emissions growth for long. We have now baked in at least 1.5°C of warming to the Earth through sloth and apathy.

Will we enter what I've called the *Hellocene*, a time of unchecked suffering and weather calamities where the world unspools like a disaster film? If images like dolphin deaths don't move us, will the "emergence of heat and humidity too severe for human tolerance" move the world to act? Heat like this is coming—on rare days some coastal tropical regions already kiss deadly combinations of heat and humidity. Wet-bulb temperatures of 35°C (95°F) mark the upper physiological limit of human survival, when our bodies can no longer cool themselves, even by sweating. Imagine an extended heat wave where people stuck outside die from the heat—like whirling dolphins—*in the shade*.

We've been warned. NASA scientist James Hansen gave landmark testimony to a U.S. Senate committee in 1988 that brimmed with evidence of climate change. More than thirty-five years ago, he concluded, "The greenhouse effect has been detected, and it is changing our climate now." Viewing the climate carnage of 2023 and the lack of action since 1988, Hansen was even stronger: "We are damned fools."

We are, and we should have listened.

Thousands of youth activists who organize global climate strikes listened. So did the sixteen youths—now grade school to college—who won an initial ruling in the case of *Held v. State of Montana*. The landmark decision struck down two Montana laws that prohibited courts and agencies from considering the climate consequences of proposed projects. The judge presiding over the case determined that Montana's constitution, which mandates the right to "a clean and healthful environment," also applies to climate.

One of the plaintiffs, Grace Gibson-Snyder, listened early in her life: "I was about 5 years old. And my best friend is from the Marshall Islands. And we heard that because of climate change, the Marshall Islands would be underwater within 50 years or so. And so we made posters that said 'Save the Marshall Islands' and hung them up around our neighborhood. I remember spelling 'ocean' O-S-H-I-N."

Gibson-Snyder retains the same fire today. When we spoke, she was a twenty-year-old college sophomore and likely global affairs major interested in climate policy. I asked what motivated her. She said, "Climate change will last our whole lifetimes. Here in Montana we see the glaciers melting and the wildfire smoke. It's oppressive; it hurts to breathe. Your eyes burn for weeks.

"I don't really have a choice in speaking out about this because I don't have a choice in experiencing it," she said. "Maybe I'll be here for another eighty years. That's a long time to see this crisis unfold if nothing's done about it. My generation will see the fallout—good or bad—of the decisions that people make now and that people have been making for fifty years."

She knows transitions take time. "We aren't asking for Montana to stop using fossil fuels tomorrow," Gibson-Snyder said. "We know how reliant Montana is upon extractive industries, and you can extrapolate that to the rest of the nation and the world. You can't just abandon these industries and the people who rely upon them. Still, there has to be a transition away from fossil fuels."

I saw a photo recently. Gibson-Snyder stood smiling—fist held high—surrounded by supporters outside the Lewis and Clark County Courthouse. The case had already consumed four years of her life. I asked what keeps her going. "I won't say that I always operate in a place of hope," she said. "But anytime someone gets involved, I feel a little bit of relief because it reminds me that climate change is not a problem that I have to solve by myself."

At the end of our last day together, Ayan Santos Fleischmann and I finally spy the Mamirauá research station—a two-story houseboat roped to riverside trees. We spot a sloth, too, and drift in to track its progress. It isn't high in the cathedral forest, it's *swimming*—head up like a life-guard—using a stroke you might expect of a sloth—a crawl. With caimans in sight, we're pulling for the sloth as it takes f o r e v e r to reach river's edge—parroting the slow pace of climate action—and finally ascends the flooded *várzea* forest.

Fleischmann and I exhale when the sloth is safe, push up the channel, and rope our boat to the floating research station at river's edge. Climbing to the second-story balcony, I peer into the forest canopy for the final time this trip.

I began my log of atmospheric restoration in a Roman cathedral and I end it in an Amazonian one. I envision the army of restorers standing on scaffolding to cleanse the sky in the frescoes of the Sistine Chapel. I celebrate the army of people rising to restore the actual atmosphere. I've dedicated the past decade of my career to this quixotic idea, and I share it in hope.

Restoring the atmosphere in a single generation is a promise, a duty, and an act of moral repair. I know that living to see methane drop to preindustrial levels is neither likely nor easy. If my dream were easy, it wouldn't be a dream.

I recall Vittoria Cimino's vision and her sense of urgency—in cathedral time—when she said, "We act for the following decade and hope the following decade will act for the decade after that."

To "act" for the Vatican meant cleaning centuries of grime, dust, and candle soot from the *Last Judgment* frescoes to reveal the breathtaking blue sky concealed beneath—an atmosphere restored for generations.

I lie on the research boat that night as the Amazon ebbs below me. Hammocks creak—gently marking time.

Stars shine through the window, a handful of the millions burning above. I see them blur, turning into flames. Each star becomes one of the millions of power plants, pilot lights, and fossil fires flickering over the Earth—each flame to be extinguished by someone's hand.

I cup a candle and exhale.

ACKNOWLEDGMENTS

Science is an alliance. I thank my colleagues in the Global Carbon Project (and beyond) for your dedication and decades of research, in particular executive director Pep Canadell, our heart and soul. I am equally indebted to the interviewees—thank you for sharing your passion and knowledge.

I thank the John Simon Guggenheim Memorial Foundation and Stanford's Doerr School and Center for Advanced Study in the Behavioral Sciences (CASBS) for sabbatical support that allowed me to start this project. Margaret Levi, Sally Schroeder, Mike Gaetani, Roberta Katz, and others made this book possible. Our CASBS Climate group included Michael Albertus, Giulia Baroni, Mario Biagioli, Michael Brownstein, Marco Casari, Alta Charo, David Ciepley, Nicole Ellison, Anita Hardon, Michael Hiscox, Bob Keohane, Paula Moya, Camilo Perez Bustillo, Cat Ramirez, Leif Wenar, and more.

My Stanford writing group has been a source of friendship, encouragement, and critique for years. I thank Russ Carpenter, Liz Carlisle (*Lentil Underground, Grain by Grain,* and *Healing Grounds*), Shannon Switzer Swanson, and Sara Michas-Martin. Members Richard Nevle (*The Paradise Notebooks*), Lauren Oakes (*In Search of the Canary Tree* and forthcoming *Treekeepers*), and Emily Polk (*Communicating Global to Local Resiliency*) have been there from the beginning. Seeing your books appear month by month—chapter by chapter—has been a revelation.

ACKNOWLEDGMENTS

Agent Max Brockman warned me that writing the book proposal would take almost as much time as writing the book. His guidance was invaluable. Editors Christopher Richards and Josephine Greywoode of Simon & Schuster and Penguin Random House UK, respectively, fielded countless questions and did much more than "edit" the manuscript. Thank you. The same holds for Joie Asuquo of Simon & Schuster and for professional editor and beloved sister-in-law Helen Graves.

Many people reviewed chapters for me. They include Yannai Kashtan, Leona Neftaliem, Peter Pellitier, Celina Scott-Buechler, Leif Wenar, Eric Lebel, Mark Shwartz, Phoebe Sweet, Michael Mullany, Sally, Rob, and Will Jackson, plus my brother, Ken, who reminded me (often) that his Colorado pipes freeze faster than mine when the power goes out. Cooper Rinzler of Breakthrough Energy provided feedback and ideas for the chapter on green steel.

Anders Ahlström and Julika Wolf hosted my trip to Scandinavia, including legs to Luleå to visit SSAB, to Jokkmokk in Swedish Lapland to meet Mats Karström, and to Sarek National Park to hike and pick cloudberries.

Jennifer Jacquet of New York University pointed me to the *Harvard Business Review* article by Bill Sells of Johns Manville on denialism in the asbestos industry.

I thank my Rice University engineering friends, including Keith Brewer and Jeff Thomas of Chapter 3, who give no slack. Ever.

For help with communications, I thank Leah Poulos and others at Press Shop PR and the Stanford communications team, including Rob Jordan, Josie Garthwaite, Chris Black, Mark Golden, and Amy Adams. Various people helped organize my visits and interviews; I acknowledge Marco Maggi of the Vatican; Margareta Rönnqvist, Viktoria Karsberg, and Mia Widell of SSAB; Wendy Martens Leja and Katie Kramon of Impossible Foods; and Erica Spooner of ADM. And although carbon offsets are imperfect, I offset emissions from all my travel associated with writing and researching this book.

ACKNOWLEDGMENTS

For the patience of the young, including students, postdocs, staff, and sons who continue to teach me. To my parents, who might not support every conclusion in this book but supported me nonetheless. And to my late father-in-law, John Graves (*Goodbye to a River*)—may you be jumping tarpon in celestial seas.

NOTES

Prologue

xiv **Sherry Rehman, Pakistan's climate minister, said** L. Sands, "Pakistan Floods: One Third of Country Is Under Water—Minister," BBC, August 30, 2022; https://www.bbc.com/news/world-europe-62712301.

xv **Rehman calls "climate colonialism"** "Sherry Slams 'Climate Colonialism' at Madrid Moot: 'Climate negotiations need to factor into the needs of low polluters who are paying the cost for decades of fossil-fuel development by rich countries,'" *The News*, December 1, 2021; https://www.thenews.com.pk/print/913181-sherry-slams-climate-colonialism-at-madrid-moot.

Introduction

xvii **Rome's hottest temperature ever** Ian Livingston and Kasha Patel, "Rome Hits Highest Temperatures on Record as Heat Wave Sweeps Europe," *Washington Post*, June 28, 2022; https://www.washingtonpost.com/climate-environment/2022/06/28/record-heat-europe-italy-rome-scandinavia/.

xvii **in a record-breaking summer where sixty thousand Europeans will die of heat** J. Ballester et al., "Heat-Related Mortality in

Europe During the Summer of 2022," *Nature Medicine* 29 (2023): 1857–66; https://doi.org/10.1038/s41591-023-02419-z.

xviii **Restoring the atmosphere—my dream as a climate scientist** R.B. Jackson and J. Salzman, "Pursuing Geoengineering for Atmospheric Restoration," *Issues in Science and Technology* 26 (2010): 67–76.

xviii **The United Nations named the 2020s the "Decade on Restoration"** https://www.decadeonrestoration.org.

xix **Vittoria Cimino's office discovered that parts of the frescoes were turning powdery white** "The Sistine Chapel Twenty Years Later: New Breath, New Light," October 30 to 31, 2014, Vatican Museums; Nicole Winfield, "Vatican: Air Pollution Is Changing the Frescoes in the Sistine Chapel," *Talking Points Memo*, October 31, 2014; https://talkingpointsmemo.com/news/vatican-admits-sistine-chapel-frescoes-whitened.

xx **trillion tons** I use the "short scale" definition of "trillion" common in the United States and Britain—10^{12} (a 1 followed by 12 zeroes) or a billion billion; in many other countries a trillion, on the "long scale," is a million times larger: 10^{18}; https://en.wikipedia.org/wiki/Trillion.

xx **Carbon dioxide is responsible for about half of recent warming . . . Methane has caused another third of recent warming** For the decade of the 2010s, carbon dioxide was responsible for ~0.75°C of global surface warming compared to the second half of the nineteenth century (1850 to 1900). Methane's contribution was two-thirds that of carbon dioxide, or 0.5°C warming over the same period. Other gases warmed the Earth, too, particularly nitrous oxide (almost 0.1°C), with another 0.1°C warming coming from halogenated gases—refrigerants such as chlorofluorocarbons (CFCs) that cause the ozone hole, and their replacement hydrofluorocarbons (HFCs). "IPCC 2021: Summary for Policymakers," in *Climate Change 2021: The Physical Science Basis. Contribu-*

tion of Working Group I to the Sixth Assessment Report of the Intergovernmental Panel on Climate Change (V. Masson-Delmotte et al., eds.).

xx **Its concentration has doubled** Throughout this book I use the term "atmospheric concentration" for accessibility, in place of the more strictly correct term "molar mixing ratios."

xxi **And methane concentrations are accelerating faster** Methane concentrations rose 15 parts per billion (ppb), 17 ppb, and 14 ppb in 2020, 2021, and 2022, respectively. Those increases were the second, first, and fourth largest since the NOAA (National Oceanic and Atmospheric Administration) methane time series began in 1983. Lan X., K. W. Thoning, and E. J. Dlugokencky, "Trends in Globally-Averaged CH_4, N_2O, and SF_6 Determined from NOAA Global Monitoring Laboratory Measurements," version 2023-04; https://doi.org/10.15138/P8XG-AA10.

xxi **Global Methane Pledge** "Participants joining the Pledge agree to take voluntary actions to contribute to a collective effort to reduce global methane emissions at least 30 percent from 2020 levels by 2030"; https://www.globalmethanepledge.org.

xxi **More than a hundred and fifty nations** As of January of 2024, the Pledge has 155 national signatories; https://www.globalmethanepledge.org.

xxi **have joined the United States and European Union** https://www.whitehouse.gov/briefing-room/statements-releases/2021/09/18/joint-us-eu-press-release-on-the-global-methane-pledge/.

xxii **particulate pollution from coal and cars kills more than a hundred thousand Americans a year** S. K. Thakrar et al., "Reducing Mortality from Air Pollution in the United States by Targeting Specific Emissions Sources," *Environmental Science and Technology Letters* 7 (2020): 639–45, https://doi.org/10.1021/acs.estlett.0c00424; see also F. Caiazzo et al., "Air Pollution and

Early Deaths in the United States. Part I: Quantifying the Impact of Major Sectors in 2005," *Atmospheric Environment* 79 (2013): 198–208, https://doi.org/10.1016/j.atmosenv.2013.05.081.

xxii **far more than are murdered, die in traffic accidents, and drown** U.S. Centers for Disease Control, deaths in 2020 from murders (24,576), traffic accidents (40,698), and drowning (4,000); https://www.cdc.gov/nchs/fastats/homicide.htm, https://www.cdc.gov/nchs/fastats/accidental-injury.htm, https://www.cdc.gov/drowning/facts/index.html.

xxii **ten million senseless deaths a year** K. Vohra et al., "2021 Global Mortality from Outdoor Fine Particle Pollution Generated by Fossil Fuel Combustion: Results from GEOS-Chem," *Environmental Research* 195 (2021): 110754; https://doi.org/10.1016/j.envres.2021.110754.

xxii **96 percent since the phaseout of leaded gasoline** "The median concentration of lead in the blood of children between the ages of 1 and 5 years dropped from 15 µg/dL in 1976–1980 to 0.6 µg/dL in 2017–2018, a decrease of 96%"; https://www.epa.gov/americaschildrenenvironment/biomonitoring-lead.

xxii **$2.5 trillion per year** P. L. Tsai, T. H. Hatfield, "Global Benefits from the Phaseout of Leaded Fuel," *Journal of Environmental Health* 74 (2011): 8–15; https://www.jstor.org/stable/26329321.

xxii **Protection of the ozone shield through the safeguards of the Montreal Protocol** A. van Djik et al., "Skin Cancer Risks Avoided by the Montreal Protocol—Worldwide Modeling Integrating Coupled Climate-Chemistry Models with a Risk Model for UV," *Photochemistry and Photobiology* 89 (2013): 234–46 and see also https://doi.org/10.1111/j.1751-1097.2012.01223.x; see also S. Madronich et al., "Estimation of Skin and Ocular Damage Avoided in the United States Through Implementation of the Montreal Protocol on Substances That Deplete the Ozone Layer," *ACS Earth and*

Space Chemistry 5 (2021): 1876–88, https://doi.org/10.1021/acs earthspacechem.1c00183.

xxii **U.S. Clean Air Act currently saves hundreds of thousands of lives** "The 1990 Clean Air Act Amendments programs are projected to result in a net improvement in U.S. economic growth and the economic welfare of American households. Our central benefits estimate exceeds costs by a factor of more than 30 to one, and the high benefits estimate exceeds costs by 90 times"; https://www .epa.gov/clean-air-act-overview/benefits-and-costs-clean-air-act -1990-2020-second-prospective-study.

xxii **Absent intervention, elevated carbon dioxide concentrations in our air will remain for tens of thousands of years** D. Archer et al., "Atmospheric Lifetime of Fossil Fuel Carbon Dioxide," *Annual Review of Earth and Planetary Sciences* 37 (2009): 117– 34; https://doi.org/10.1146/annurev.earth.031208.100206.

xxiv **Doing so has cooled the planet and saved millions of lives** S. O. Andersen, M. L. Halberstadt, and N. Borgford-Parnell, "Stratospheric Ozone, Global Warming, and the Principle of Unintended Consequences—an Ongoing Science and Policy Success Story," *Journal of the Air & Waste Management Association* 63 (2013): 607–47; https://doi.org/10.1080/10962247.2013.791349.

Chapter 1: Fair Shares

3 **We live in a time when the top 1 percent of the world's population contributes more fossil carbon emissions than half the people on Earth** L. Chancel, "Global Carbon Inequality over 1990–2019," *Nature Sustainability* 5 (2022): 931–38; https:// www.nature.com/articles/s41893-022-00955-z.

3 **I led a recent analysis of health, well-being, and energy use across 140 countries** R. B. Jackson et al., "Human Wellbeing and

Per Capita Energy Use," *Ecosphere* 13 (2022): 3978; https://doi
.org/10.1002/ecs2.3978.

4 **"How much energy powers a good life?"** "How much energy
powers a good life? Less than you're using, says a new report,"
All Things Considered, NPR, April 12, 2022; https://one.npr
.org/?sharedMediaId=1092045712:1092414700.

4 **"World Doesn't Need More Energy to End Poverty"** D. R.
Baker, "World Doesn't Need More Energy to End Poverty, Study
Says," Bloomberg News, April 12, 2022; https://www.bloomberg
.com/news/articles/2022-04-12/world-doesn-t-need-more-ener
gy-to-end-poverty-study-says.

4 **"Where the Energy Link to Well-Being Starts Fraying"**
B. Geman, "Where the Energy Link to Well-Being Starts Fraying,"
Axios, April 12, 2022; https://www.axios.com/2022/04/12/ener
gy-well-being-consumption.

5 **with almost one passenger vehicle per person** Hedges & Company, https://hedgescompany.com/blog/2021/06/how-many
-cars-are-there-in-the-world/#:~:text=There%20are%201.4
46%20billion%20cars%20in%20the%20world%20in%202022,
https://en.wikipedia.org/wiki/List_of_countries_by_vehicles
_per_capita.

6 **more mining of rare metals such as gallium and indium** U.S.
Geological Survey, "Critical Mineral Commodities in Renewable
Energy" (2019); https://www.usgs.gov/media/images/critical
-mineral-commodities-renewable-energy.

6 **a recent opinion piece in the *New York Times*,** Rob Jackson
and Pep Canadell, "What's 'Fair' When It Comes to Carbon Emissions?," *New York Times*, December 4, 2019; https://www.ny
times.com/2019/12/04/opinion/global-climate-change.html.

6 **a tireless advocate for poorer communities in the southern
United States** Justine Calma, "Sewage Is Still 'America's Dirty
Secret,'" *The Verge*, November 13, 2020; https://www.theverge

.com/21564385/sewage-america-dirty-secret-catherine-flowers
-interview-wastewater.

7 "Environmental justice" is defined by the U.S. Environmental
Protection Agency https://www.epa.gov/environmentaljustice.

7 Residents fall prey to diseases that are normally found only
in the poorest tropical countries P. Hotez, "Hookworm and
Poverty," *Annals of the New York Academy of Sciences* 1136
(2008): 38–44; https://nyaspubs.onlinelibrary.wiley.com/doi/pdf
/10.1196/annals.1425.000.

8 more than a third of residents they tested in Lowndes County
had hookworm M. L. McKenna et al., "Human Intestinal Parasite
Burden and Poor Sanitation in Rural Alabama," *American Journal
of Tropical Medicine and Hygiene* 97 (2017): 1623–28; https://
doi.org/10.4269/ajtmh.17-0396.

8 Hookworms affect more than half a billion people worldwide
Centers for Disease Control and Prevention; https://www.cdc
.gov/parasites/hookworm/index.html.

8 Scientists call this process "global worming" A. J. Blum and P. J.
Hotez, "Global 'Worming': Climate Change and Its Projected General Impact on Human Helminth Infections," *PLOS One* 12 (2018):
e0006370; https://doi.org/10.1371/journal.pntd.0006370.

9 A sacrifice zone "Sacrifice Zones 101," Climate Reality Project;
https://www.climaterealityproject.org/sacrifice-zones.

9 people of color comprise 45 percent of residents M. Mascar-
enhas, R. Grattet, and K. Mege, "Toxic Waste and Race in Twenty-
First Century America," *Environment and Society* 12 (2021);
https://doi.org/10.3167/ares.2021.120107.

9 People living within six miles of a coal or natural gas power plant
are also far more likely to be poor and of color J. Declet-Barreto
and A. A. Rosenberg, "Environmental Justice and Power Plant
Emissions in the Regional Greenhouse Gas Initiative States," *PLoS
One* 17 (2022); https://doi.org/10.1371/journal.pone.0271026.

9 Such facilities are disproportionately located near where immigrants live L. Laurian and R. Funderburg, "Environmental Justice in France? A Spatio-Temporal Analysis of Incinerator Location," *Journal of Environmental Planning and Management* 57 (2013): 424-46; https://doi.org/10.1080/09640568.2012.749395.

9 nitrogen dioxide (NO_2) pollution generated by fossil fuel combustion "Nitrogen Dioxide (NO_2) Is a Gas That Is Mainly Produced During the Combustion of Fossil Fuels," U.K. Department for Environment, Food, and Rural Affairs, 2023 National Statistics Nitrogen Dioxide (NO_2); https://www.gov.uk/government/statistics/air-quality-statistics/ntrogen-dioxide.

9 the national average G. Mitchell and D. Dorling, "An Environmental Justice Analysis of British Air Quality," *Environmental Planning A: Economy and Space* 35 (2003): 909–29; https://doi.org/10.1068/a35240.

10 two times more drenched days with three or more inches of rain than it used to K. E. Kunkel, "Change in Heavy Precipitation," in "Impacts, Risks, and Adaptation in the United States: The Fourth National Climate Assessment," Volume II, 2018, Figure 19.3; https://data.globalchange.gov/report/nca4/chapter/southeast/figure/change-in-heavy-precipitation.

11 The National Oceanic and Atmospheric Administration (NOAA) has been collecting data on billion-dollar weather disasters in the United States NOAA National Centers for Environmental Information (NCEI), "U.S. Billion-Dollar Weather and Climate Disasters, 2024"; https://www.ncei.noaa.gov/access/billions/.

11 Half of all U.S. households facing energy insecurity are African American R. D. Bullard, "Sacrifice Zones: the Front Lines of Toxic Chemical Exposure in the United States," *Environmental Health Perspectives* 119 (2011): A266; https://doi.org/10.1289/ehp.119-a266.

Chapter 2: Home on the Range

13 **our homes and buildings contribute a third or more of all greenhouse gas emissions in the United States and Europe** *U.S. Environmental Protection Agency 2021 Sources of Greenhouse Gas Emissions*, https://www.epa.gov/ghgemissions /sources-greenhouse-gas-emissions; for Europe, see J. Rankin, "EU Urged to Ratchet Up Green Energy Standards for Buildings," *Guardian*, December 14, 2021, https://www.theguardian.com /environment/2021/dec/14/eu-urged-to-rachet-up-green-energy -standards-for-buildings.

13 **Gas use ranges from 40 percent of homes in Australia** "Gas Stoves and Damp Houses Increase Aussie Asthma Rates," *University of Queensland News*, April 16, 2018; https://www.uq.edu.au /news/article/2018/04/gas-stoves-and-damp-houses-increase -aussie-asthma-rates.

13 **and roughly half of U.S. homes** U.S. Energy Information Administration 2023, *Natural Gas Explained*; https://www.eia.gov /energyexplained/natural-gas/use-of-natural-gas.php.

13 **to nearly all homes in the Netherlands** M. Smedley, "Dutch Seek to Reduce Gas Use in Homes," *Natural Gas World* (2016); https:// www.naturalgasworld.com/dutch-look-at-reducing-gas-use-in -homes-34803.

14 **roughly 370 million tons of methane pollution from human activities each year** M. Saunois et al., "The Global Methane Budget 2000–2017," *Earth System Science Data* 12 (2020): 1561– 1623, https://doi.org/10.5194/essd-2019-128; R. B. Jackson et al., "Increasing Anthropogenic Methane Emissions Arise Equally from Agricultural and Fossil Fuel Sources," *Environmental Research Letters* 15 (2020): 071002, https://iopscience.iop.org /article/10.1088/1748-9326/ab9ed2.

14 **38 billion tons of fossil carbon dioxide pollution globally in 2023 alone** P. Friedlingstein et al., "Global Carbon Budget 2023," *Earth System Science Data* 15 (2023): 5301–69; https://doi .org/10.5194/essd-15-5301-2023.

15 **Referring to carbon dioxide, which yielded the highest temperature, Foote wrote** E. Foote, "Circumstances Affecting the Heat of the Sun's Rays," *American Journal of Science* (1856): 382–83; https://publicdomainreview.org/collection/first-paper -to-link-co2-and-global-warming-by-eunice-foote-1856/.

15 **Arrhenius even predicted a rate of warming** S. Arrhenius, *Worlds in the Making: The Evolution of the Universe* (New York and London: Harper and Brothers Publishers, 1908); https:// archive.org/details/worldsinmakinge01arrhgoog/page/n1 /mode/2up.

15 **Recent work suggests a doubling of atmospheric CO_2** R. Lindsey, "How Much Will Earth Warm if Carbon Dioxide Doubles Pre-industrial Levels?," National Oceanic and Atmospheric Administration, 2014; https://www.climate.gov/news-features/climate -qa/how-much-will-earth-warm-if-carbon-dioxide-doubles -pre-industrial-levels.

15 **Exxon's internal estimate from their models of warming from the early 1980s** G. Supran, S. Rahmstorf, and N. Oreskes, "Assessing ExxonMobil's Global Warming Projections," *Science* 379 (2023): 6628; https://www.science.org/doi/10.1126/science.abk0063.

15 **More than 60 percent of U.S. households use gas for heating or for cooking** U.S. Energy Information Administration Residential Energy Consumption Survey (RECS), 2020; https://www.eia .gov/consumption/residential/.

17 **our previous research has shown that storage-tank water heaters** E. D. Lebel et al., "Quantifying Methane Emissions from Natural Gas Water Heaters," *Environmental Science and Technology* 54 (2020): 5737–45; https://doi.org/10.1021/acs.est.9b07189.

18 **In our first gas stove study** E. D. Lebel et al., "Methane and NO$_x$ Emissions from Natural Gas Stoves, Cooktops and Ovens in Residential Homes," *Environmental Science and Technology* 56 (2022): 2529–39; https://doi.org/10.1021/acs.est.1c04707.

18 **Most people's exposure to benzene** American Cancer Society; https://www.cancer.org/cancer/cancer-causes/benzene.html.

18 **On average, the gasoline sold in the United States** Gasoline Standards: Gasoline Mobile Source Air Toxics, U.S. Environmental Protection Agency, "The National Benzene Content of Gasoline Today Is About 1.0 Vol%"; https://www.epa.gov/gasoline-standards /gasoline-mobile-source-air-toxics.

19 **WHO also concludes that for cancer effects** World Health Organization, 2010, *WHO Guidelines for Indoor Air Quality: Selected Pollutants*; https://apps.who.int/iris/handle/10665/260127.

19 **(surveys show most people don't)** N. L. Nagda et al., "Prevalence, Use, and Effectiveness of Range-Exhaust Fans," *Environment International* 15 (1989): 615–20, https://doi.org/10.1016/0160 -4120(89)90083-4; see also H. Zhao et al., "Factors Impacting Range Hood Use in California Houses and Low-Income Apartments," *International Journal of Environmental Research and Public Health* 17 (2020): 8870, https://doi.org/10.3390/ijerph17238870.

20 **Nitrogen dioxide is a respiratory irritant** Environmental Protection Agency, "Asthma Triggers: Gain Control"; https://www.epa.gov /asthma/asthma-triggers-gain-control#nitro: "Studies show a connection between breathing elevated short-term NO$_2$ concentrations, and increased visits to emergency departments and hospital admissions for respiratory issues, especially asthma."

20 **The U.S. EPA's national ambient air quality standard (NAAQS) for nitrogen dioxide outdoors** "Primary National Ambient Air Quality Standards (NAAQS) for Nitrogen Dioxide," 2023; https:// www.epa.gov/no2-pollution/primary-national-ambient-air-quali ty-standards-naaqs-nitrogen-dioxide.

21 **A 2013 summary of forty-one studies** W. Lin, B. Brunekreef, and U. Gehring, "Meta-analysis of the Effects of Indoor Nitrogen Dioxide and Gas Cooking on Asthma and Wheeze in Children," *International Journal of Epidemiology* 42 (2013): 1724–37; https://doi.org/10.1093/ije/dyt150.

21 **My group's work showed that breathing NO2 pollution** Y. Kashtan et al., "Nitrogen Dioxide Exposure, Health Outcomes, and Associated Demographic Disparities Due to Gas and Propane Combustion by U.S. Stoves," *Science Advances* (2024).

21 **as documented in the association's proceedings** Proceedings of the Natural Gas Association of America, second annual meeting in Joplin, Missouri, May 21–23, 1907, pages 192–93 and 198–201; https://books.google.com/books?id=WIRCAQAAIAAJ&printsec=front cover&source=gbs_ge_summary_r&cad=0#v=onepage&q&f=false.

22 **and those of other scientists** For example, B. C. Singer, et al., "Pollutant Concentrations and Emission Rates from Natural Gas Cooking Burners Without and With Range Hood Exhaust in Nine California Homes," *Building and Environment* 122 (2017): 215–29; https://doi.org/10.1016/j.buildenv.2017.06.021.

23 **rather than the common hydrofluorocarbon refrigerants** https://en.wikipedia.org/wiki/R-410A.

23 **The recent U.S. High-Efficiency Electric Home Rebate Act includes up to $14,000** A. Picchi, "Biden's inflation law offers up to $14,000 for home upgrades. Here's how to qualify," CBS News, September 13, 2022; https://www.cbsnews.com/news/inflation -reduction-act-joe-biden-climate-energy-home-upgrades/.

Chapter 3: Planet of the Cows

26 **Food production produces a third of greenhouse gas emissions globally** M. Crippa et al., Food systems are re-

sponsible for a third of global anthropogenic GHG emissions, approximately 18 billion metric tons of carbon-dioxide equivalents a year. *Nature Food* 2 (2021): 198–209; https://doi.org/10.1038/s43016-021-00225-9.

26 **Beef production, for instance, currently drives 80 percent of Amazon deforestation** *Yale School of Forestry and Environmental Studies: Global Forest Atlas 2020.*

26 **Methane contributes another third of greenhouse gas emissions from agriculture** M. Saunois et al., "The Global Methane Budget 2000–2017," *Earth System Science Data* 12 (2020): 1561–623, https://doi.org/10.5194/essd-2019-128; see also R. B. Jackson et al., "Increasing Anthropogenic Methane Emissions Arise Equally from Agricultural and Fossil Fuel Sources," *Environmental Research Letters* 15 (2020): 071002, https://iopscience.iop.org/article/10.1088/1748-9326/ab9ed2.

26 **which comes out to a few hundred pounds a year** K. A. Johnson and D. E. Johnson, "Methane Emissions from Cattle," *Journal of Animal Science* 73 (1995): 2483–92; https://pubmed.ncbi.nlm.nih.gov/8567486/.

26 **a 2019 *New Yorker* article** Tad Friend, "Can a Burger Help Save Climate Change?," *New Yorker*, September 23, 2019; https://www.newyorker.com/magazine/2019/09/30/can-a-burger-help-solve-climate-change.

26 **In 2022 the United States imported 3.5 billion pounds of beef and exported 3.4 billion pounds** USDA Economic Research Service, "Livestock and Meat International Trade Data," U.S. Department of Agriculture, 2023; https://www.ers.usda.gov/data-products/livestock-and-meat-international-trade-data.

27 **My colleagues and I recently estimated** C. Hong et al., "Global and Regional Drivers of Land-Use Emissions in 1961–2017," *Nature* 589 (2021): 554–61; https://www.nature.com/articles/s41586-020-03138-y.

27 **The average American eats four times more beef a year** Fifty-eight pounds of beef a year in the U.S. and fourteen pounds for an average person globally. "Meat Consumption," "OECD-FAO Agricultural Output 2021"; https://data.oecd.org/agroutput /meat-consumption.htm.

27 **the three hundred million cows now slaughtered yearly** Food and Agriculture Organization of the United Nations, "Livestock Systems"; http://www.fao.org/livestock-systems/global-distributions /cattle/en/.

27 **a quarter of Earth's ice-free land . . . and one-third of global croplands** "Twenty-six percent of the Planet's ice-free land is used for livestock grazing and 33 percent of croplands are used for livestock feed production," Food and Agriculture Organization of the United Nations, "Livestock and Landscapes"; http://www.fao.org /3/ar591e/ar591e.pdf.

27 **Pat Brown, founder and former CEO of Impossible Foods** My quotes from Pat Brown come from two sources. Most of them come from a personal interview on April 26, 2021. A few others come from a seminar he gave to the Stanford Woods Institute for the Environment, March 11, 2021 ("A Conversation with Pat Brown"; https://earth .stanford.edu/events/conversation-pat-brown-impossible-foods).

28 **can be found in Burger Kings** https://impossiblefoods.com /locations.

30 **U.S. farmers grow seventy-five million acres of soybeans** "Factsheet," 2015, USDA.gov; https://www.usda.gov/sites/default /files/documents/coexistence-soybeans-factsheet.pdf.

30 **energy transfer between organisms is only about 10 percent efficient** J. B. Reece et al., "Trophic Pyramids," in *Campbell Biology*, 10th edition, Chapter 55.3 (Saddle River, NJ: Pearson Publishing, 2011).

30 **Cornell ecologist David Pimentel wrote** "US Could Feed 800 Million People with Grain That Livestock Eat, Cornell Ecologist

Advises Animal Scientists," *Cornell Chronicle*, August 7, 1997; https://news.cornell.edu/stories/1997/08/us-could-feed-800 -million-people-grain-livestock-eat.

30 **A pound of soybeans takes about two hundred gallons of water** Betty Hallock, "To make a burger, first you need 660 gallons of water . . . ," *Los Angeles Times*, January 27, 2014; https://www .latimes.com/food/dailydish/la-dd-gallons-of-water-to-make-a -burger-20140124-story.html.

31 **The authors of the study wrote** B. D. Richter et al., Water Scarcity and Fish Imperilment Driven by Beef Production," *Nature Sustainability* 3 (2020): 319–28; https://doi.org/10.1038 /s41893-020-0483-z.

31 **Lakes Mead and Powell** A. Kaur, "Changes Needed to Save Second-Largest U.S. Reservoir, Experts Say," *Washington Post*, February 18, 2023; https://www.washingtonpost.com/climate -environment/2023/02/18/changes-needed-save-second-largest -us-reservoir-experts-say/.

31 **Lake Mead is being emptied by a megadrought turbocharged by climate change** A. P. Williams, B. I. Cook, and J. E. Smerdon, "Rapid Intensification of the Emerging Southwestern North American Megadrought in 2020–2021," *Nature Climate Change* 12 (2022): 232–34; https://doi.org/10.1038/s41558-022-01290-z.

32 **"cover crops"** "Cover Crops and Crop Rotation," U.S. Department of Agriculture; https://www.usda.gov/peoples-garden/soil -health/cover-crops-crop-rotation.

33 **"antibiotic resistance in humans"** M. J. Martin, S. E. Thottathil, and T. B. Newman, "Antibiotics Overuse in Animal Agriculture: A Call to Action for Health Care Providers," *American Journal of Public Health* 105 (2015): 2409–10; https://ajph.aphapublica tions.org/doi/full/10.2105/AJPH.2015.302870.

34 **A comparison from the Harvard Medical School** Emily Gelsomin, MLA, RD, LDN, "Impossible and Beyond: How Healthy Are

These Meatless Burgers?," Harvard Health Publishing, January 24, 2022; https://www.health.harvard.edu/blog/impossible-and-beyond-how-healthy-are-these-meatless-burgers-2019081517448.

34 **Another recent analysis, in the journal *Food Science and Human Wellness*** B. M. Bohrer, "An Investigation of the Formulation and Nutritional Composition of Modern Meat Analogue Products," *Food Science and Human Wellness* 8 (2019): 320–29; https://doi.org/10.1016/j.fshw.2019.11.006.

34 **Many studies have found that plant-based diets** A. Satija and F. B. Hu, "Plant-Based Diets and Cardiovascular Health," *Trends in Cardiovascular Medicine* 28 (2018): 437–41; https://doi.org/10.1016/j.tcm.2018.02.004.

34 **obesity, gastrointestinal cancers, type-2 diabetes** For obesity and gastrointestinal cancers: M. W. Ewy et al., "Plant-Based Diet: Is It as Good as an Animal-Based Diet When It Comes to Protein?," *Current Nutrition Reports* 11 (2022): 337–46, https://link.springer.com/article/10.1007/s13668-022-00401-8; for type-2 diabetes: M. McMacken and S. Shah, "A Plant-Based Diet for the Prevention and Treatment of Type-2 Diabetes," *Journal of Geriatric Cardiology* 14 (2017): 342–54, https://www.ncbi.nlm.nih.gov/pmc/articles/PMC5466941/.

35 **Cheese consumption has doubled in the United States** Cheese accounts for largest share of per capita U.S. dairy product consumption, USDA; https://www.ers.usda.gov/data-products/chart-gallery/gallery/chart-detail/?chartId=103984.

37 **a record fire season across the western United States and Canada** https://en.wikipedia.org/wiki/2020_Western_United_States_wildfire_season#cite_note-NOAAbillion-2.

37 **Portland shattered its all-time temperature record** "Monday's record-setting temperatures broke Sunday's record-setting high of 112 degrees. Sunday's high had broken the 108-degree record set Saturday, which broke the previous high of 107, first set in

1965." https://www.oregonlive.com/weather/2021/06/portland
-records-all-time-high-temperature-of-113-setting-new-record
-for-third-day-in-a-row.html; June 29, 2021.

37 **set the highest temperature ever recorded in Canada at
121°F** "Lytton, British Columbia, hit 121 degrees on Tuesday—
the highest temperature ever recorded in Canada." https://www
.cnn.com/2021/06/29/americas/canada-heat-wave-deaths
/index.html.

37 **(almost 50°C)** "More than 230 deaths have been reported in
British Columbia since Friday as a historic heat wave brought
record-high temperatures, officials said Tuesday. The province's
chief coroner called it an "unprecedented time." https://www.cnn
.com/2021/06/29/americas/canada-heat-wave-deaths/index.html

37 **Lytton burned to the ground the next day** "'From what I can see
in town there's only a few houses left so it'll be a total rebuild,' Jan
Polderman said." https://bc.ctvnews.ca/it-ll-be-a-total-rebuild
-lytton-b-c-mayor-describes-work-underway-and-next-steps
-1.5494709.

38 **"so large and generating so much energy and extreme heat that
it's changing the weather"** https://www.cnn.com/2021/07/20
/weather/us-western-wildfires-tuesday/index.html.

38 **Everyone I know in the western United States or British Columbia has a recent fire story** Rob Jackson, "The West Is Going
Up in Flames," *Scientific American*, August 29, 2018; https://
blogs.scientificamerican.com/observations/the-west-is-going-up
-in-flames/.

39 **It was once called the "Timber Capital of the Nation"** "In 1939
there were only 37 sawmills operating in the county, but by 1947
this number increased to 278, most of them engaged in cutting
Douglas fir lumber. The harvesting and processing of this 'green
gold' earned Roseburg the title of 'Timber Capital of the Nation'";
https://www.cityofroseburg.org/visitors/history.

41 In 2020, Burger King launched a "reduced-methane Whopper" https://www.rbi.com/English/sustainability/respon 'Timber / Nation.' " sible-sourcing/beef-cow-methane/.

41 They launched the campaign with a regrettable, surrealist video https://www.youtube.com/watch?v=G-P7VceH9F8.

42 some of the first studies evaluating whether seaweed food supplements could reduce methane emissions B. Roque et al., "Inclusion of *Asparagopsis Armata* in Lactating Dairy Cows' Diet Reduces Enteric Methane Emission by Over 50 Percent," *Journal of Cleaner Production* 234 (2019): 132–38, https://doi.org/10.1016/j .jclepro.2019.06.193. Earlier Australian studies showed beneficial effects in laboratory studies, including R. D. Kinley et al., "The Red Macroalgae *Asparagopsis taxiformis* Is a Potent Natural Antimethanogenic That Reduces Methane Production During In Vitro Fermentation with Rumen Fluid," *Animal Production Science* 56 (2016): 282–89; https://doi.org/10.1071/AN15576.

45 Here, at last, is the gas quantification system S. A. Gunter and M. R. Beck, "Measuring the Respiratory Gas Exchange by Grazing Cattle Using an Automated, Open-Circuit Gas Quantification System," *Translational Animal Science* 2 (2018): 11–18; https://doi .org/10.1093/tas/txx009.

46 Early studies examined See A. N. Hristov et al., "An Inhibitor Persistently Decreased Enteric Methane Emission from Dairy Cows with No Negative Effect on Milk Production," *Proceedings of the National Academy of Sciences U.S.A.* 112 (2015): 10663–68 and references therein; https://doi.org/10.1073/pnas.1504124112.

46 3-NOP 3-nitrooxypropanol; https://en.wikipedia.org/wiki/3-Ni trooxypropanol.

46 leaves of the lemongrass plant M. F. Vázquez-Carrillo et al., "Effect of *Cymbopogon citratus* on Enteric Methane Emission, Nutrients Digestibility, and Energy Partition in Growing Beef Cattle,"

Archivos Latinoamericanos De Producción Animal 31 (2023): 207–12; https://doi.org/10.53588/alpa.310537.

46 **the enzyme that microbes use in the last step of producing methane** Methyl-coenzyme M reductase; https://en.wikipedia .org/wiki/Coenzyme-B_sulfoethylthiotransferase.

46 **small amounts of *Asparagopsis armata* seaweed . . . cut methane emissions by two-thirds** B. M. Roque et al., "Inclusion of *Asparagopsis armata* in Lactating Dairy Cows' Diet Reduces Enteric Methane Emission by Over 50 Percent," *Journal of Cleaner Production* 234 (2019): 132–38; https://doi.org/10.1016/j.jcle pro.2019.06.193.

47 **Remarkably, methane emissions dropped 82 percent** B. M. Roque et al., "Red Seaweed (*Asparagopsis taxiformis*) Supplementation Reduces Enteric Methane by Over 80 Percent in Beef Steers," *PLOS One* 16 (2021): e0247820; https://doi.org/10.1371 /journal.pone.0247820.

49 **Bromoform (in seaweeds) is not the only chemical additive** For information on other chemicals and approaches to reduce methane emissions from cows, including vaccines, see J. L. Black, T. M. Davison, and I. Box, "Methane Emissions from Ruminants in Australia: Mitigation Potential and Applicability of Mitigation Strategies," *Animals* 11 (2021): 951; https://doi.org/10.3390 /ani11040951.

49 **3-NOP has been trademarked by the Dutch company Royal DSM** "Bovaer: Farm-wise, climate-friendly," DSM press release; https://www.dsm.com/corporate/markets/animal-feed /minimizing-methane-from-cattle.html.

50 **"short-lived climate pollutants"** "S.B. 1383: SB-1383: methane emissions: dairy and livestock: organic waste: landfills"; https://leginfo.legislature.ca.gov/faces/billNavClient.xhtml?bill _id=201520160SB1383.

50 **"carbon neutral" butter** https://www.foodnavigator.com/Article/2021/03/17/Fonterra-talks-carbon-neutral-dairy; independent third-party verification provided by *Toitū Envirocare* (https://www.toitu.co.nz).

50 **("Simply Milk")** "New Zealand's First Carbonzero Milk," 2020, https://www.fonterra.com/nz/en/our-stories/articles/new-zealands-first-carbonzero-milk.html.

50 **consumers are contributing to carbon offsets** Olivia Wannan, "New 'Carbon-Neutral' Milk Offsets Its Greenhouse Gas Emissions," 2020, https://www.stuff.co.nz/environment/climate-news/122113648/new-carbonneutral-milk-offsets-its-greenhouse-emissions.

50 **New Zealand approved 3-NOP in August of 2023** "EPA Approves First Methane Inhibitor in New Zealand," Environmental Protection Authority, August 10, 2023; https://www.epa.govt.nz/news-and-alerts/latest-news/epa-approves-first-methane-inhibitor-in-new-zealand/.

50 **Fonterra already signed an agreement** "Fonterra Joins Forces with DSM to Lower Carbon Footprint," January 29, 2021; https://www.fonterra.com/nz/en/our-stories/media/fonterra-joins-forces-with-dsm-to-lower-carbon-footprint.html.

50 **Blue Ocean Barns** https://www.blueoceanbarns.com.

51 **Mootral** https://mootral.com.

51 **based on research by none other than Ermias Kebreab** B. M. Roque et al., "Effect of Mootral—A Garlic- and Citrus-Extract-Based Feed Additive—on Enteric Methane Emissions in Feedlot Cattle," *Translational Animal Science* 3 (2019): 1383–88; https://doi.org/10.1093/tas/txz133.

Chapter 4: REVved Up

52 **Zero Motorcycles** https://zeromotorcycles.com.

52 **CZU Lightning Complex fires** CAL FIRE, "CZU Lightning Complex (Including Warnella Fire)" "Confirmed damage to property: 1,490 structures destroyed"; https://www.fire.ca.gov/incidents /2020/8/16/czu-lightning-complex-including-warnella-fire/.

53 **Joby Aviation** "Joby Aviation Announces Closing of Business Combination with Reinvent Technology Partners to Become Publicly Traded Company," August 10, 2021, "Reid Hoffman, LinkedIn Co-Founder and Co-Lead Director of RTP, added, 'With its advanced technology, we believe Joby is "Tesla meets Uber in the air" and the clear leader in the eVTOL and aerial ridesharing space." https://www.jobyaviation.com/news/joby-aviation-closing -business-combination-with-rtp/.

54 **Transportation, including travel by road, rail, air, and water** Environmental Protection Agency, "Sources of Greenhouse Gas Emissions," "Transportation (27 percent of 2020 greenhouse gas emissions for the United States)—The transportation sector generates the largest share of U.S. greenhouse gas emissions. Greenhouse gas emissions from transportation primarily come from burning fossil fuel for our cars, trucks, ships, trains, and planes. Over 90 percent of the fuel used for transportation is petroleum based, which includes primarily gasoline and diesel." https:// www.epa.gov/ghgemissions/sources-greenhouse-gas-emissions.

54 **average age for cars** "Average Age of Cars and Light Trucks in the U.S. Rises to 12.1 Years, Accelerated by COVID-19," IHS Markit, June 14, 2021; https://www.spglobal.com/mobility/en /research-analysis/average-age-of-cars-and-light-trucks-in-the -us-rises.html.

54 **and motorcycles** Eric R. Teoh, "Motorcycles Registered in the United States, 2002–18," Insurance Institute for Highway Safety, January 2019, "The average age of registered motorcycles has increased from 8.7 years in 2002 to 12.0 years in 2018 (Table 4b). Half of motorcycles registered in 2018 were at least 11 years old." https://www.iihs.org/api/datastoredocument/bibliography/2181.

54 **seven years for buses** "Average Age of Urban Transit Vehicles," 6.4 years for the average age of transit buses in the United States in 2021, U.S. Bureau of Transportation Statistics; https://www.bts .gov/content/average-age-urban-transit-vehicles-years.

54 **thirteen years for commercial airplanes** "Average Age of Aircraft 2018," Bureau of Transportation Statistics; https://www.bts .gov/content/average-age-aircraft.

54 **twenty-eight years for train locomotives** Statista, "Average Age of the North American Locomotive Fleet from 2014 to 2020," average age of 28.1 years for locomotives in the United States in 2020; https://www.statista.com/statistics/580894/north-american -locomotives-average-age/.

54 **BNSF proudly states** "BNSF Locomotives"; https://www.bnsf .com/about-bnsf/virtual-train-tour/locomotive.html.

54 *Car & Driver* **notes** Hearst Autos Research, "How Many Miles Does a Car Last?"; https://www.caranddriver.com/research/a32 758625/how-many-miles-does-a-car-last/.

54 **And each time someone fills up at a gas station** "Greenhouse Gas Emissions from a Typical Passenger Vehicle"; https://www .epa.gov/greenvehicles/greenhouse-gas-emissions-typical -passenger-vehicle; a 20-gallon fill-up therefore releases 177,000 g or ~392 pounds of CO_2.

54 **four and a half tons of carbon pollution a year** "A typical passenger vehicle emits about 4.6 metric tons of carbon dioxide per year. This assumes the average gasoline vehicle on the road today has a fuel economy of about 22.0 miles per gallon and drives approx-

imately 11,500 miles per year. Every gallon of gasoline burned creates about 8,887 grams of CO_2. This number can vary based on a vehicle's fuel, fuel economy, and the number of miles driven per year." https://www.epa.gov/greenvehicles/greenhouse-gas -emissions-typical-passenger-vehicle.

54   **an average price of \$3.96 per gallon** "Weekly Retail Gasoline and Diesel Prices," U.S. Energy Information Administration (EIA), September 2023; https://www.eia.gov/dnav/pet/pet_pri_gnd _dcus_nus_m.htm.

54   **tailpipe emissions kill fifty thousand Americans a year** S. Anenberg et al., "A global snapshot of the air pollution–related health impacts of transportation sector emissions in 2010 and 2015," George Washington University Milken Institute School of Public Health, International Council on Clean Transportation, University of Colorado, Boulder, 2019; https://theicct.org/sites/default/ files/Global_health_impacts_transport_emissions_2010- 2015_20190226_1.pdf.

55   **People in Germany and the United Kingdom are three times more likely to ride mass transit regularly than in the United States** "2023 Transportation," International Comparisons; https:// internationalcomparisons.org/environmental/transportation/.

55   **one of the lowest rates worldwide** D. Jones, "Bike Commuting Statistics: 74 Cycling to Work Stats for 2024," Discerning Cyclist, January 2, 2024, https://discerningcyclist.com/bike-commuting- statistics/; for Sweden and the Netherlands, see S. Yanatma, "Cycling in Europe: Which Countries and Cities Are the Most and Least Bicycle-Friendly?," October 10, 2023, https://www .euronews.com/next/2023/09/19/cycling-in-europe-which -countries-and-cities-are-the-most-and-least-bicycle-friendly.

55   **Studies show that bicycle commuting reduces stress** I. Avila-Palencia et al., "The Relationship Between Bicycle Commuting and Perceived Stress: A Cross-Sectional Study," *BMJ Open: Epidemi-*

ology Research 7 (2017): e013542; https://bmjopen.bmj.com /content/7/6/e013542.

55 **four in ten Americans still breathe dangerous air** "American Lung Association 2023 State of the Air"; https://www.lung.org /research/sota.

55 **More Americans die from car pollution** ~53,000 vs. 34,000, respectively. F. Caiazzo et al., "Air Pollution and Early Deaths in the United States. Part I: Quantifying the Impact of Major Sectors in 2005," *Atmospheric Environment* 79 (2013): 198–208; https://www.sciencedirect.com/science/article/abs/pii/S135 2231013004548.

55 **strokes** R. Maheswaran and P. Elliott, "Stroke Mortality Associated with Living Near Main Roads in England and Wales: A Geographical Study," *Stroke* 34 (2004); https://www.ahajournals.org /doi/full/10.1161/01.STR.0000101750.77547.11.

55 **heart attacks** A. Peters et al., "Exposure to Traffic and the Onset of Myocardial Infarction," *New England Journal of Medicine* 351 (2004): 1721–30; https://www.nejm.org/doi/full/10.1056/NEJ Moa040203.

55 **dementia** H. Chen et al., "Living Near Major Roads and the Incidence of Dementia, Parkinson's Disease and Multiple Sclerosis: A Population-Based Cohort Study," *Lancet* 389 (2017): 718–26; https://doi.org/10.1016/S0140-6736(16)32399-6.

55 **asthma and "lifetime wheeze"** D. Y. Jung et al., "Effect of Traffic-Related Air Pollution on Allergic Disease: Results of the Children's Health and Environmental Research," *Allergy, Asthma and Immunology Research* 7 (2015): 359–66, http://dx.doi.org/10.4168 /aair.2015.7.4.359; G. Bowatte et al., "Traffic-Related Air Pollution Exposure Over a 5-Year Period Is Associated with Increased Risk of Asthma and Poor Lung Function in Middle Age," *European Respiratory Journal* 50 (2017): 1602357, https://erj .ersjournals.com/content/50/4/1602357.

55 **As shown in England and Wales** R. Maheswaran and P. Elliott, "Stroke Mortality Associated with Living Near Main Roads in England and Wales: A Geographical Study," *Stroke* 34 (2003): 2776–80; https://doi.org/10.1161/01.STR.0000101750.77547.11.

56 **According to the U.S. EPA** "Near Roadway Air Pollution and Health: Frequently Asked Questions"; https://www.epa.gov/sites/production/files/2015-11/documents/420f14044_0.pdf.

56 **So are people of color** "People of Color Are 3.7 Times More Likely than White People to Live in a County with Three Failing Grades," "The State of the Air, 2023," American Lung Association; https://www.lung.org/research/sota/key-findings.

56 **The EPA concludes** Environmental Protection Agency, "Research on Near Roadway and Other Near Source Air Pollution"; https://www.epa.gov/air-research/research-near-roadway-and-other-near-source-air-pollution.

57 **in the United States** "Motorcycles," National Highway Traffic Safety Administration, Motorcycles comprise 3 percent of registered vehicles; https://crashstats.nhtsa.dot.gov/Api/Public/ViewPublication/812785.

57 **Canada** In 2019, 736,216 registered motorcycles and mopeds out of 35,742,412 total registered vehicles. "Vehicle registrations by type of vehicle," Statistics Canada Table 23-10-0067-01; https://www150.statcan.gc.ca/t1/tbl1/en/tv.action?pid=2310006701.

57 **and the United Kingdom** Statista, "Volume of Motorcycles and Cars Registered in Great Britain Between 2000 and 2022," "About 1.32 million motorcycles and almost 31.9 million cars were registered in Great Britain at the end of 2020"; https://www.statista.com/statistics/312594/motorcycle-and-car-registrations-in-the-united-kingdom/.

57 **Most households own at least one motorcycle or motorbike** WorldAtlas, "Countries with the Highest Motorbike Usage," India (47 percent), China (60 percent), Indonesia (85 percent),

and Thailand (87 percent); https://www.worldatlas.com/articles /countries-that-ride-motorbikes.html.

58 *Mythbusters* **did tests of their own** "*MythBusters* Asks: Are Motorcycles Greener Than Cars?," *Los Angeles Times*, September 28, 2011; https://latimesblogs.latimes.com/greenspace /2011/09/mythbusters-motorcycle-emissions.html.

58 **a Danish company, Vestas** https://www.vestas.com/en/about /this-is-vestas.

59 **Sales of all EVs reached fourteen million in 2023, more than one in six of all cars sold globally** "Zero-emissions Vehicles Factbook," BloombergNEF, December 2023; https://assets .bbhub.io/professional/sites/24/2023-COP28-ZEV-Factbookpdf.

59 **The car companies that sold the most EVs in 2023** Insideevs, August 7, 2023; https://insideevs.com/news/680475 /world-top-ev-oem-sales-2023h1/.

59 **Tesla's market value of $800 billion** Julie Pinkerton, *U.S. News and World Report*, October 16, 2023 (Tesla's Market Capitalization on October 24th, 2023: $801 billion; Toyota's: $244 billion); https://money.usnews.com/investing/articles/the-10-most -valuable-auto-companies-in-the-world.

59 **Ford plans to invest $22 billion** "The Ford Electric Vehicle Strategy: What You Need to Know," Ford EV commitments, May 19, 2021; https://media.ford.com/content/fordmedia/fna /us/en/news/2021/05/19/the-ford-electric-vehicle-strategy --what-you-need-to-know.html.

60 **GM's financial commitment to EVs** June 16, 2021; https:// investor.gm.com/news-releases/news-release-details/gm-will -boost-ev-and-av-investments-35-billion-through-2025/.

60 **CEO Mary Barra said** Jameson Dow, Electrek, November 19, 2020; https://electrek.co/2020/11/19/gm-30-new-evs-by-2025 -7-billion-to-evs/.

60 **spoke to our students about GM's plans to electrify its fleet**
SIEPR associates meeting with Mary Barra, chief executive officer
of General Motors, January 13, 2022; https://www.youtube.com
/watch?v=yuofY5o6muU.

60 **On LinkedIn, Barra wrote** Michael Wayland, CNBC, January 28,
2021; https://www.cnbc.com/2021/01/28/general-motors-plans
-to-exclusively-offer-electric-vehicles-by-2035.html.

61 **Europe, every EV generates less than one-third** P. Mock and G.
Bieker, "Fact Sheet: European Union CO_2 standards for new passenger cars and vans: life-cycle greenhouse gas emissions," International Council on Clean Transportation, July 21, 2021; https://
theicct.org/publication/european-union-co2-standards-for
-new-passenger-cars-and-vans-life-cycle-greenhouse-gas
-emissions/.

61 **According to U.S. Department of Energy data** "Lifetime Carbon Emission of Electric Vehicles vs Gasoline Cars," Reuters, July
7, 2021; https://www.reuters.com/business/autos-transporta
tion/lifetime-carbon-emissions-electric-vehicles-vs-gasoline-cars
-2021-06-29/.

62 **"Hydrogen Road Tour '08"** http://www.hydrogenroadtour.com.

62 **hydrogen refueling stations** U.S. Department of Energy;
https://afdc.energy.gov/fuels/hydrogen_locations.html#/find
/nearest?fuel=HY, https://www.glpautogas.info/en/hydrogen-sta
tions-united-states.html.

62 **fewer than nineteen thousand globally in 2023 and only three
thousand in the United States** R. Parkes, "Passenger Cars Accounted for 80% of Hydrogen Vehicle Sales in 2022—But That
Share Is Shrinking," Hydrogen Insight, January 27, 2023; https://
www.hydrogeninsight.com/transport/passenger-cars-accounted
-for-80-of-hydrogen-vehicle-sales-in-2022-but-that-share-is
-shrinking/2-1-1394857.

62 automakers sold 14 million EVs globally in 2023 N. Carey, "Global Electric Car Sales Rose 31% in 2023," Reuters, January 10, 2024; https://www.reuters.com/business/autos-transportation /global-electric-car-sales-rose-31-2023-rho-motion-2024-01-11/.

62 it's ten to thirty times worse for climate N. Warwick et al., "Atmospheric Implications of Increased Hydrogen Use, 2022; https://www.gov.uk/government/publications/atmospheric -implications-of-increased-hydrogen-use.

63 in response to COVID-19 "Japan Creates $19bn Green Fund to Push Hydrogen Planes and Carbon Recycling"; https://newson japan.com/html/newsdesk/article/129182.php.

66 (Before throwing the prince under the bus) "The Prince Albert II of Monaco Foundation works to protect the environment and promote sustainable development. Active at the international level, the Foundation focuses its actions on three main areas: limiting the effects of climate change and promoting renewable energies, preserving biodiversity, managing water resources and combating desertification." https://www.fpa2.org/en/index.

68 1 percent of the industrial hydrogen manufactured today International Energy Administration, "The Future of Hydrogen," "While less than 0.1% of global dedicated hydrogen production today comes from water electrolysis, with declining costs for renewable electricity, in particular from solar PV and wind, there is growing interest in electrolytic hydrogen." https://www.iea.org /reports/the-future-of-hydrogen.

68 most hydrogen made today International Energy Administration, "The Future of Hydrogen," "Natural gas is currently the primary source of hydrogen production, accounting for around three quarters of the annual global dedicated hydrogen production of around 70 million tonnes. This accounts for about 6% of global natural gas use. Gas is followed by coal, due to its dominant role in China." https://www.iea.org/reports/the-future-of-hydrogen.

68 **It's generated by steam-methane reforming** U.S. Energy Information Administration, "Hydrogen Explained," "Commercial hydrogen producers and petroleum refineries use steam-methane reforming to separate hydrogen atoms from carbon atoms in methane (CH4). In steam-methane reforming, high-temperature steam (1,300°F to 1,800°F) under 3–25 bar pressure (1 bar = 14.5 pounds per square inch) reacts with methane in the presence of a catalyst." https://www.eia.gov/energyexplained/hydrogen/production-of-hydrogen.php.

68 **larger than the fossil carbon emissions of France and the United Kingdom combined** International Energy Administration, "Hydrogen"; https://www.iea.org/fuels-and-technologies/hydrogen.

68 **Most of this hydrogen is used by industry** "Hydrogen Explained," U.S. Energy Information Administration; https://www.eia.gov/energyexplained/hydrogen/use-of-hydrogen.php.

68 **Green hydrogen was recently three times more expensive** "As of 2020 green hydrogen costs between $2.50–6.80 per kilogram and turquoise hydrogen $1.40–2.40/kg or blue hydrogen $1.40–2.40/kg compared with high-carbon grey hydrogen at $1–1.80/kg." https://en.wikipedia.org/wiki/Hydrogen_economy#Costs.

68 **the International Renewable Energy Agency projects** December 17, 2020, "Hydrogen produced with renewable electricity could compete on costs with fossil fuel alternatives by 2030, according to a new report from the International Renewable Energy Agency (IRENA) published today. A combination of falling costs for solar and wind power, improved performance as well as economies of scale for electrolysers could make it possible." https://www.irena.org/News/pressreleases/2020/Dec/Making-Green-Hydrogen-a-Cost-Competitive-Climate-Solution.

68 **Over a century, hydrogen's global warming potential is eleven times greater than that of carbon dioxide** "Our estimate of the hydrogen GWP for a 100 year time horizon is 11 ± 5, which is more than 100% larger than previously published calculations. Approximately one third of the GWP arises due to the stratospheric response, which was not considered in previous studies." N. Warwick et al., "Atmospheric Implications of Increased Hydrogen Use," 2022; https://www.gov.uk/government/publications /atmospheric-implications-of-increased-hydrogen-use.

68 **One international estimate suggests** International Energy Administration, "The Future of Hydrogen," "Producing all of today's dedicated hydrogen output from electricity would result in an electricity demand of 3,600 TWh, more than the total annual electricity generation of the European Union." https://www.iea.org /reports/the-future-of-hydrogen.

69 **Pound for pound, it has almost three times** U.S. Department of Energy, "Hydrogen Basics," "The energy in 2.2 pounds (1 kilogram) of hydrogen gas is about the same as the energy in 1 gallon (6.2 pounds, 2.8 kilograms) of gasoline." https://afdc.energy.gov /fuels/hydrogen_basics.html.

70 **"*Solar Impulse* was built not to transport passengers"** https://bertrandpiccard.com/exploration/solar-impulse.

Chapter 5: Stop the Steel

72 **Steel companies worldwide** Global steel production in 2020 was 1.86 billion metric tons. The global average carbon intensity of steel production that year was 1.85 metric tons of CO_2 per ton of steel. World Steel Association, *Steel Statistical Yearbook 2021*; https://www.worldsteel.org.

72 **The industry is responsible for a towering 11 percent of global carbon dioxide emissions** C. Swalec, "These 553 steel plants are responsible for 9% of global CO2 emissions," Carbon Brief, 2021 guest post; https://www.carbonbrief.org/guest-post-these-553 -steel-plants-are-responsible-for-9-of-global-co2-emissions/.

73 **Global steel manufacturing burns more than a billion tons of metallurgical coal** "World Steel Association 2022 Fact Sheet: steel and raw materials"; https://worldsteel.org/wp-content/uploads /Fact-sheet-raw-materials-2023.pdf.

73 **more coal than the United States and European Union use altogether** For the United States: U.S. Energy Information Administration, "Coal Explained," https://www.eia.gov/energyexplained/ coal/use-of-coal.php; for the E.U.: "International Energy Administration 2022 Global Coal Consumption 2020–2023," https:// www.iea.org/data-and-statistics/charts/global-coal-consumption -2020-2023.

74 **to strip the extra oxygen found in iron ores** Because iron ores in minerals such as magnetite (Fe_3O_4) and hematite (Fe_2O_3) contain much more oxygen than is desired for making steel, companies need something to strip out the extra oxygen.

75 **About three-quarters of current global steel production uses a blast furnace** "Today about 73% of steel is produced using the BF-BOF; 26% is produced via the EAF route," with "BF-BOF" referring to blast furnace-basic oxygen furnace and "EAF" referring to electric arc furnace. World Steel Association; https://worldsteel.org /about-steel/steel-facts/.

78 **already producing heavy-duty electric trucks made with HYBRIT steel** J. Strandhede, "World-First: Volvo Delivers Electric Trucks with Fossil-Free Steel to Customers," October 11, 2022; https://www.volvogroup.com/en/news-and-media/news/2022 /nov/news-4396618.html.

78 **the HYBRIT partners are investing** LKAB and emissions-free steel, https://www.lkab.com/en/; HYBRIT project: T. K. Blanch, "Sweden Goes for Zero-Carbon Steel," energypost.eu, December 16, 2020, https://energypost.eu/hybrit-project-sweden-goes-for-zero-carbon-steel/.

78 **national carbon prices** "Carbon Pricing Dashboard," World Bank; https://carbonpricingdashboard.worldbank.org/map_data.

78 **Sweden's carbon price rose fivefold** Government Offices of Sweden; https://government.se/government-policy/swedens-carbon-tax/swedens-carbon-tax/.

80 **Groundbreaking climate work is happening in Sweden** "Sweden levies the highest carbon tax rate in the world, at SEK 1,190 (US $126) per metric ton of CO_2. The tax is primarily levied on fossil fuels used for heating purposes and motor fuels." S. Jonsson, A. Ydstedt, and E. Asen, "Looking Back on 30 Years of Carbon Taxes in Sweden," September 23, 2020; https://taxfoundation.org/sweden-carbon-tax-revenue-greenhouse-gas-emissions/.

80 **four of every five cars Norwegians purchased were already electric** P. Johnson, "This Is the Norway—Nation Hits Record EV Share in 2022 on Its Way to Ending Gas Car Sales," Electrek, January 2 2023; https://electrek.co/2023/01/02/norway-hits-record-ev-share-in-2022/.

80 **median internal corporate carbon price** "Putting a Price on Carbon," *CDP Report 2021*; https://www.cdp.net/en/research/global-reports/putting-a-price-on-carbon.

80 **Sweden** Swedish Climate Policy Council; https://www.klimatpolitiskaradet.se/wp-content/uploads/2019/09/climatepolicycouncilreport2.pdf.

80 **Finland** Statistics Finland; https://www.stat.fi/til/khki/2019/khki_2019_2020-05-28_tie_001_en.html.

80 **U.S. greenhouse gas emissions** U.S. EPA; https://www.epa.gov /climate-indicators/climate-change-indicators-us-greenhouse -gas-emissions.

81 **Wind turbines are three-quarters steel** U.S. Geological Survey, "What Materials Are Used to Make Wind Turbines?" "Wind Turbines Are Predominantly Made of Steel (66–79% of Total Turbine Mass)"; https://www.usgs.gov/faqs/what-materials-are-used-make -wind-turbines.

82 **("plyscrapers")** Z. Gorvett, "Plyscrapers: The Rise of the Wooden Skyscraper," BBC, October 31, 2017; https://www.bbc.com/future /article/20171026-the-rise-of-skyscrapers-made-of-wood.

82 **Business Roundtable** https://www.businessroundtable.org.

Chapter 6: Pipe Dreams

84 **Hundreds of thousands of gas leaks pepper the local pipeline networks** Z. D. Weller, S. P. Hamburg, and J. C. von Fischer, "A National Estimate of Methane Leakage from Pipeline Mains in Natural Gas Local Distribution Systems," *Environmental Science and Technology* 54 (2020): 8958–67; https://doi.org/10.1021 /acs.est.0c00437.

84 **the EPA underestimates urban methane emissions by a factor of at least three to five** Various studies all point to such underestimates for urban methane emissions: K. McKain et al., "Methane Emissions from Natural Gas Infrastructure and Use in the Urban Region of Boston, Massachusetts," *Proceedings of the National Academy of Sciences USA* 112 (2015): 1941–46, https://doi .org/10.1073/pnas.1416261112; G. Plant et al., "Evaluating Urban Methane Emissions from Space Using TROPOMI Methane and Car-

bon Monoxide Observations," *Remote Sensing of Environment* 268 (2022): 112756, https://doi.org/10.1016/j.rse.2021.112756; see also Z. D. Weller et al., "A National Estimate of Methane Leakage from Pipeline Mains in Natural Gas Local Distribution Systems," *Environmental Science and Technology* 54 (2020): 8958–67, https://doi.org/10.1021/acs.est.0c00437.

86 **Some cast-iron pipes in use today date back to the Civil War generals** "Cast and wrought iron pipelines were originally constructed to transport manufactured gas beginning in the 1870s and 1880s, with cast iron becoming more popular in the early 1900s." "Cast and wrought iron inventory," Pipeline and Hazardous Materials Safety Administration; https://www.phmsa.dot.gov/data-and-statistics/pipeline-replacement/cast-and-wrought-iron-inventory.

87 **we detected explosive methane concentrations inside manholes** R. B. Jackson et al., "Natural Gas Pipeline Leaks Across Washington, D.C.," *Environmental Science and Technology* 48 (2014): 2051–58, https://pubs.acs.org/doi/10.1021/es404474x; C. Joyce, "About 6,000 Natural Gas Leaks Found in D.C.'s Aging Pipes," National Public Radio, January 16, 2014, https://www.npr.org/2014/01/16/262911327/aging-pipes-in-d-c-create-about-6-000-natural-gas-leaks.

88 **The number one predictor of a natural gas leak along streets in Boston** N. G. Phillips et al., "Mapping Urban Pipeline Leaks: Methane Leaks Across Boston," *Environmental Pollution* 173 (2013): 1–4; https://doi.org/10.1016/j.envpol.2012.11.003.

88 **a gas-related apartment explosion** "NTSB Faults Utility's Equipment in Fatal Maryland Gas Explosion," Associated Press, April 23, 2019; https://www.nbcnews.com/news/us-news/ntsb-faults-utility-s-equipment-fatal-maryland-gas-explosion-n997801.

90 **Then Representative (and later Senator) Ed Markey wrote a letter** J. Mulholland, "Researchers Map Boston's Natural Gas

Leaks," December 6, 2012, https://www.govtech.com/dc/articles /researchers-map-bostons-natural-gas-leaks.html; see also "America Pays for Gas Leaks," August 1, 2013, https://www.markey.sen ate.gov/imo/media/doc/documents/markey_lost_gas_report .pdf.

90 **our Boston study** N. G. Phillips et al., "Mapping Urban Pipeline Leaks: Methane Leaks Across Boston," *Environmental Pollution* 173 (2013): 1–4; https://doi.org/10.1016/j.envpol.2012.11 .003.

90 **Massachusetts passed an accelerated pipeline repair and replacement bill** Massachusetts 2104, Chapter 149, An Act Relative to Natural Gas Leaks; https://malegislature.gov/Laws/Session-Laws/Acts/2014/Chapter149.

91 **We achieved the same outcome a year later** R. B. Jackson et al., "Natural Gas Pipeline Leaks Across Washington, D.C.," *Environmental Science and Technology* 48 (2014): 2051–58; https:// pubs.acs.org/doi/10.1021/es404474x.

91 **PROJECT*pipes*** https://dcpsc.org/Newsroom/HotTopics/Infra structure-Enhancements/PROJECTpipes.aspx#.

91 **The District's Public Service Commission** https://dcpsc.org /Newsroom/HotTopics/Infrastructure-Enhancements/PRO JECTpipes.aspx.

91 **supply gas to more than seventy-five million households in the United States alone** Sixty-one percent of U.S. homes use gas. "The majority of U.S. households used natural gas in 2020," U.S. Energy Information Administration; https://www.eia.gov /todayinenergy/detail.php?id=55940. The total number of U.S. households is 124 million. "Quick Facts, United States," U.S. Census Bureau; https://www.census.gov/quickfacts/fact/table/US /HSD410221.

91 **The state of Victoria in Australia** T. Cosoleto, "Cooking with Gas No More as Victoria Heads to Net Zero," *Canberra*

Times, July 27, 2023; https://www.canberratimes.com.au /story/8287005/cooking-with-gas-no-more-as-victoria-heads -for-net-zero/.

91 **Vancouver, Canada, has required** E. Chung, "Why Oil and Gas Heating Bans for New Homes Are a Growing Trend," CBC News, January 30, 2022; https://www.cbc.ca/news/science/bans-fossil-fuel-heating-homes-1.6327113.

92 **More than a hundred municipalities in the United States have passed similar requirements** "Zero Emission Building Ordinances," Building Decarbonization Coalition; https://buildingdecarb.org/zeb-ordinances.

92 **They have already paid homeowners in some areas to buy new electric appliances** https://docs.google.com /presentation/d/1A6C_-wtp7nC491m4-P3f0HA6UX1ElhAh /edit#slide=id.p11.

92 **For instance, PG&E has to replace twelve hundred miles of older plastic pipes** J. Van Derbeken, "PG&E to Replace 1,200 Miles of Plastic Gas Pipe," SFGate, October 14, 2011; https:// www.sfgate.com/news/article/PG-E-to-replace-1-200-miles-of -plastic-gas-pipe-2327084.php.

92 **PG&E asked the CA Public Utilities Commission for permission** K. Abraham, "East Campus May Become California's Largest Electrification Project," *Cal State University Monterey Bay News*, September 7, 2022; https://csumb.edu/news /news-listing/east-campus-may-become-californias-largest -electrification-project/.

93 **the Southern California Gas Corporation (SoCalGas) that serves Los Angeles and Santa Barbara sued the California Energy Commission** T. DiChristopher, "SoCalGas Sues California Energy Commission to Block 'Anti–Natural Gas Policy,'" S&P Global, August 5, 2020; https://www.spglobal.com /marketintelligence/en/news-insights/latest-news-headlines

/socalgas-sues-california-energy-commission-to-block-anti-natu
ral-gas-policy-59758122.

94 **the new station would further harm a community whose
residents already bear a legacy of contamination** "Health Impact Assessment of a Proposal Natural Gas Compressor Station
in Weymouth, MA," Massachusetts Department of Health, Massachusetts Department of Environmental Protection, and Metropolitan Area Planning Council, January 2019; https://www.mass
.gov/doc/health-impact-assessment-weymouth-proposed-natural
-gas-compressor-station-executive-summary/download.

95 **The Weymouth Compressor Station had three "unplanned
gas releases"** M. Wasser, "Weymouth Compressor Reports Another 'Unplanned' Gas Release. Third Time in 8 Months," WBUR
.org, April 6, 2021; https://www.wbur.org/news/2021/04/06
/weymouth-compressor-gas-release.

96 **A quick search on social media showed road closures** "Manhole Explosions Shut Down Charles Circle," WBZ, March 24, 2020;
https://www.youtube.com/watch?v=eB6Pqcw0rI8.

97 **Recent estimates suggest more than six hundred thousand
gas leaks plague U.S. streets alone** Z. D. Weller, S. P. Hamburg,
and J. C. von Fischer, "A National Estimate of Methane Leakage
from Pipeline Mains in Natural Gas Local Distribution Systems,"
Environmental Science and Technology 54 (2020): 8958–67;
https://doi.org/10.1021/acs.est.0c00437.

97 **Massachusetts eliminated all major coal-fired power . . .
in-state electricity generation** U.S. Energy Information Administration, November 16, 2023, "In 2001, coal fueled almost
three-tenths of Massachusetts' electricity net generation, but since
mid-2017 there has been no utility-scale (1 megawatt or larger)
coal-fired electricity generation in the state." "Natural gas fueled
two-thirds of Massachusetts' total in-state electricity net generation in 2022," U.S. Energy Information Administration, Massa-

chusetts State Profile and Energy Estimates; https://www.eia.gov
/state/analysis.php?sid=MA.

97 **The D.C. Public Service Commission** "Since inception, WGL has
remediated 28.6 miles of main and 5,188 services through PRO-
JECT*pipes* phases 1 and 2 as of December 31, 2021"; https://dcpsc
.org/Newsroom/HotTopics/Infrastructure-Enhancements
/PROJECTpipes.aspx.

Chapter 7: CFC Repair

98 **Swap methane's four hydrogens for two chlorine and two fluorine atoms** https://en.wikipedia.org/wiki/Dichlorodifluoro-
methane.

98 **CFCs 11 and 12 are super-powered greenhouse gases** "Greenhouse Gas Protocol: Global Warming Potential Values (4,660
for CFC-11 and 10,200 for CFC-12, based on the 5th Assessment Report [AR5])"; https://ghgprotocol.org/sites/default
/files/ghgp/Global-Warming-Potential-Values%20%28Feb%20
16%202016%29_1.pdf.

99 **Stratospheric ozone is Earth's sunscreen** "Ozone Layer Protection," U.S. EPA; https://www.epa.gov/ozone-layer-protection.

99 **The U.S. Environmental Protection Agency estimates the protocol prevented** "With full implementation of the Montreal Protocol, the U.S. Environmental Protection Agency (EPA) estimates
that Americans born between 1890 and 2100 are expected to
avoid 443 million cases of skin cancer, approximately 2.3 million
skin cancer deaths, and more than 63 million cases of cataracts,
with even greater benefits worldwide." "The Montreal Protocol on
Substances That Deplete the Ozone Layer," U.S. Department of
State, Office of Environmental Quality; https://www.state.gov/key

-topics-office-of-environmental-quality-and-transboundary
-issues/the-montreal-protocol-on-substances-that-deplete-the
-ozone-laycr/.

103 **CFC-11 concentrations weren't dropping as fast as they should have been** S. A. Montzka et al., "An Unexpected and Persistent Increase in Global Emissions of Ozone-Depleting CFC-11," *Nature* 557 (2018): 413–17; https://doi.org/10.1038/s41586 -018-0106-2.

103 **they concluded that the increased CFC-11 emissions likely came from eastern Asia** S. A. Montzka et al., "An Unexpected and Persistent Increase in Global Emissions of Ozone-Depleting CFC-11," *Nature* 557 (2018): 413–17; https://doi.org/10.1038 /s41586-018-0106-2.

104 **"a new source or sources of emissions from China's Shandong province after 2012"** M. F. Lunt et al., "Continued Emissions of the Ozone-Depleting Substance Carbon Tetrachloride from Eastern Asia," *Geophysical Research Letters* 45 (2018): 11,423–30; https://doi.org/10.1029/2018GL079500.

105 **The *New York Times* published an exposé** C. Buckley and H. Fountain, "In a High-Stakes Environmental Whodunit, Many Clues Point to China," *New York Times*, June 24, 2018; https:// www.nytimes.com/2018/06/24/world/asia/china-ozone-cfc.ht ml?module=inline.

106 **Matt Rigby, Steve Montzka, and others documented** S. A. Montzka et al., "A Decline in Global CFC-11 Emissions During 2018–2019," *Nature* 590 (2021): 428–32; https://doi .org/10.1038/s41586-021-03260-5.

107 **A NASA statement** "Ozone Hole Continues Shrinking in 2022, NASA and NOAA Scientists Say," NASA, October 26, 2022; https:// www.nasa.gov/esnt/2022/ozone-hole-continues-shrinking -in-2022-nasa-and-noaa-scientists-say.

107 **According to one recent study** R. Goyal et al., "Reduction in Surface Climate Change Achieved by the 1987 Montreal Protocol," *Environmental Research Letters* 14 (2019): 124041; https://iop science.iop.org/article/10.1088/1748-9326/ab4874.

108 **Its global concentration peaked at around 150 parts per trillion** "NOAA Global Monitoring Laboratory, CH_3CCl_3 (methyl-chloroform)"; https://gml.noaa.gov/aftp/data/hats/solvents/CH 3CCl3/flasks/GCMS/CH3CCL3_GCMS_flask.txt.

Chapter 8: Drawdown

Much of this chapter was first published in Greta Thunberg, ed., *The Climate Book* (Penguin Press, 2023).

111 **about three million Empire State buildings** Empire State Realty Trust, The mass of the Empire State building is ~365,000 tons; https://www.esbnyc.com/sites/default/files/esb_fact_sheet _4_9_14_4.pdf.

113 **In fact, annual global fossil carbon dioxide emissions** P. Friedlingstein et al., "Global Carbon Budget 2022," *Earth System Science Data* 14 (2022): 4811–900, https://doi.org/10.5194 /essd-14-4811-2022; R. B. Jackson et al., "Global Fossil Carbon Emissions Rebound Near Pre-COVID-19 Levels," *Environmental Research Letters* 17 (2022): 031001, https://doi .org/10.1088/1748-9326/ac55b6.

113 **One of our recent Global Carbon Project analyses** S. Fuss et al., "Moving Toward Net-Zero Emissions Requires New Alliances for Carbon Dioxide Removal," *One Earth* 3 (2020): 145–49; https:// doi.org/10.1016/j.oneear.2020.08.002.

113 **larger than the combined annual GDPs of China and the United States** $35 billion combined in 2020. "Gross domestic

product (GDP) at current prices in China and the United States from 2005 to 2020 with forecasts until 2035," Statista; https://www.statista.com/statistics/1070632/gross-domestic-product-gdp-china-us/.

114 **In 2023 there were only about forty carbon capture and storage (CCS) plants** Number of CCS plants operating today: 41 in 2023 with 26 under construction. M. Steyn et al., "2023 Status Report, Global CCS Institute"; https://status23.globalccsinstitute.com

114 **The amounts of carbon stored annually rose to twenty-nine million tons of carbon dioxide** Y. Zhang, C. Jackson, and S. Krevor, "An Estimate of the Amount of Geological CO_2 Storage over the Period of 1996–2020," *Environmental Science and Technology Letters* 9 (2022): 693–98; https://doi.org/10.1021/acs.estlett.2c00296.

114 **If all of those fossil plants complete the end of their lifetimes** D. Tong et al., "Committed Emissions from Existing Energy Infrastructure Jeopardize 1.5°C Climate Target," *Nature* 572 (2019): 373–77; https://doi.org/10.1038/s41586-019-1364-3.

114 **Agricultural activities such as plowing have released billions of tons of carbon dioxide** Agriculture soil carbon losses: J. Sanderman, T. Hengl, and G. J. Fiske, "Soil Carbon Debt of 12,000 Years of Human Land Use," *Proceedings of the National Academy of Sciences* 114 (2017): 9575–80; https://doi.org/10.1073/pnas.1706103114.

114 **Fairly optimistic estimates suggest such practices** B. W. Griscom et al., "Natural Climate Solutions," *Proceedings of the National Academy of Sciences* 114 (2017): 11645–50; https://doi.org/10.1073/pnas.1710465114.

115 **to offset anywhere near the almost forty billion metric tons** P. Friedlingstein et al., "Global Carbon Budget 2023," *Earth System Science Data* 15 (2023): 5301–69; https://doi.org/10.5194/essd-15-5301-2023.

115 **Plants . . . Rocks and industrial chemicals can also be used** UK Royal Society, *Greenhouse Gas Removal* (London: Royal Society, 2019); https://royalsociety.org/-/media/policy/projects/greenhouse-gas-removal/royal-society-greenhouse-gas-removal-report-2018.pdf.

115 **A recent U.S. National Academy of Sciences study** "The U.S. lower bound range is approximately 35 percent of the maximum potential flux; thus, the global lower bound BECCS CO2 flux potential is estimated to be 3.5-5.2 Gt/y CO2 by 2050." *Negative Emissions Technologies and Reliable Sequestration: A Research Agenda* (Washington, D.C.: The National Academies Press, 2019); https://doi.org/10.17226/25259.

116 **Another drawdown technology is enhanced weathering** D. J. Beerling et al., "Potential for Large-Scale CO_2 Removal Via Enhanced Rock Weathering with Croplands," *Nature* 583 (2020): 242–48; https://doi.org/10.1038/s41586-020-2448-9.

117 **The current cost range for direct-air capture** K. Lebling et al., "6 Things to Know About Direct Air Capture," World Resources Institute, 2021; https://www.wri.org/insights/direct-air-capture-resource-considerations-and-costs-carbon-removal.

117 **Two-thirds of global methane emissions** M. Saunois et al., "The Global Methane Budget 2000–2017," *Earth System Science Data* 12 (2020): 1561–623, https://doi.org/10.5194/essd-12-1561-2020; R. B. Jackson et al., "Increasing Anthropogenic Methane Emissions Arise Equally from Agricultural and Fossil Fuel Sources," *Environmental Research Letters* 15 (2020): 071002, https://iopscience.iop.org/article/10.1088/1748-9326/ab9ed2.

117 **If feasible at scale, methane removal** S. Abernethy et al., "Methane Removal and the Proportional Reductions in Surface Temperature and Ozone," *Philosophical Transactions of the Royal Society A* 379 (2021): 20210104; https://doi.org/10.1098/rsta.2021.0104.

118 **Insurance giant Swiss Re** "The Costs of Climate Change," Swiss Re, April 22, 2021; https://www.swissre.com/institute/research /topics-and-risk-dialogues/climate-and-natural-catastrophe-risk /expertise-publication-economics-of-climate-change.html.

119 **our recent analysis** J. C. S. Long et al., "Clean Firm Power Is the Key to California's Carbon-Free Energy Future," *Issues in Science and Technology*, March 24, 2021; https://issues.org/california -decarbonizing-power-wind-solar-nuclear-gas/.

Chapter 9: Out of Gas

120 **The world added hundreds of new coal-fired power plants in 2022** "Global Coal Plant Tracker," July 2, 2023, Global Energy Monitor; https://globalenergymonitor.org/projects/global-coal -plant-tracker/.

120 **Utilities in the United States have not built a new coal-fired power plant since 2015** Energy Information Administration, February 5, 2024; https://www.eia.gov/todayinenergy /detail.php?id=50658.

121 **Still, BECCS is relatively cheap by negative emission standards** National Academies of Sciences, Engineering and Medicine, *Negative Emissions Technologies and Reliable Sequestration: A Research Agenda* (Washington, D.C.: The National Academies Press, 2019); https://doi.org/10.17226/25259.

122 **U.S. manufacturers make more than half of the world's ethanol** https://www.statista.com/statistics/1106345/distribution-of -global-ethanol-production-by-country/.

122 **U.S. ethanol production releases 40 million tons** S. Gollakota and S. McDonald, "Carbon Dioxide Capture, Utilization, and Storage: An Emerging Economic Opportunity for Fuel Ethanol Plants in the U.S.," International Fuel Ethanol Workshop & Expo, St.

Louis, Missouri, 2013; https://www.netl.doe.gov/sites/default
/files/netl-file/Gollakota_Wed12June_Track3_FEW2013.pdf.

123 **For example, the FutureGen project** FutureGen, in particular,
is a cautionary tale of the challenges of scaling CCS technologies.
"FutureGen Fact Sheet: Carbon Dioxide Capture and Storage Project"; https://sequestration.mit.edu/tools/projects/futuregen.html.

123 **In 2009, the U.S. Department of Energy committed a billion
dollars to the project** P. Folger, "The FutureGen Carbon Capture and Sequestration Project: A Brief History and Issues for
Congress," Congressional Research Service, February 10, 2014;
https://sgp.fas.org/crs/misc/R43028.pdf.

124 **the U.S. recently announced $2.5 billion for carbon-
management projects** "Biden-Harris Administration Announces $2.5 Billion to Cut Pollution and Deliver Economic
Benefits to Communities Across the Nation," U.S. Department
of Energy, February 23, 2023; https://www.energy.gov/articles
/biden-harris-administration-announces-25-billion-cut-pollution
-and-deliver-economic.

124 **"when the [Department of Energy] established regional
partnerships"** J. T. Lytinsky et al., "The U.S. Department of Energy's Regional Carbon Sequestration Partnership Program: A
Collaborative Approach to Carbon Management," *Environment
International* 32 (2006): 128–44; https://pubmed.ncbi.nlm.nih
.gov/16054694/.

124 **"the Mount Simon formation"** S. M. Frailey, J. Damico, and
H. E. Leetaru, "Reservoir Characterization of the Mt. Simon Sandstone, Illinois, Basin, USA," *Energy Procedia* 4 (2011): 5487–94;
https://doi.org/10.1016/j.egypro.2011.02.534.

124 **Decatur was the world's first bioenergy carbon capture and
storage project** "Carbon Storage Atlas," U.S. Department of Energy;
https://netl.doe.gov/coal/carbon-storage/atlas/mgsc/phase-III
/ibdp.

125 **(Oil wells produce far more wastewater than oil)** "Challenges of Reusing Produced Water," Society of Petroleum Engineers; https://www.spe.org/en/industry/challenges-in-reusing-produced-water/.

125 **(including natural radioactive materials)** "TENORM: Oil and Gas Production Wastes," U.S. Environmental Protection Agency; https://www.epa.gov/radiation/tenorm-oil-and-gas-production-wastes.

125 **About a trillion gallons of brines . . . "Class II" injection wells** "Class II Oil and Gas Related Injection Wells," U.S. Environmental Protection Agency; https://www.epa.gov/uic/class-ii-oil-and-gas-related-injection-wells.

125 **"The United States looks like a pin cushion"** A. Lustgarten, "Injection Wells: The Poison Beneath Us," *ProPublica*, June 21, 2012; https://www.propublica.org/article/injection-wells-the-poison-beneath-us.

125 **this one from the left-leaning *High Country News*** E. Guerin, "Can More Oil Extraction Cut CO2 Emissions from Power Plants?," *High Country News*, December 9 2013; https://www.hcn.org/issues/45.21/can-more-oil-extraction-cut-co2-emissions-from-power-plants.

125 **Recent analyses suggest that EOR may sometimes help climate *a little*** R. Farajzadeh et al., "On the Sustainability of CO_2 Storage Through CO_2–Enhanced Oil Recovery," *Applied Energy* 261 (2020): 114467; https://www.sciencedirect.com/science/article/pii/S0306261919321555.

125 **most EOR operations in the United States today use CO_2 *mined from underground*** Christophe McGlade, "Can CO2-EOR Really Provide Carbon-Negative Oil?," Energy Information Administration, November 19, 2019; https://www.iea.org/commentaries/can-co2-eor-really-provide-carbon-negative-oil.

127 **Several early CCS projects in Europe were canceled** S. Akerboom et al., "Different This Time? The Prospects of CCS in the

Netherlands in the 2020s," *Frontiers in Energy Research* 9 (2021); https://doi.org/10.3389/fenrg.2021.644796.

127 **endanger their town and reduce home prices** C. F. J. Feenstra, T. Mikunda, and S. Brunsting, "What Happened in Barendrecht? Case Study on the Planned Onshore Carbon Dioxide Storage in Barendrecht, the Netherlands," Global CCS Institute, 2010; https://www.globalccsinstitute.com/archive/hub/publications/8172/barendrecht-ccs-project-case-study.pdf.

127 **A postmortem by the Global CCS Institute** C. F. J. Feenstra, T. Mikunda, and S. Brunsting, "What Happened in Barendrecht? Case Study on the Planned Onshore Carbon Dioxide Storage in Barendrecht, the Netherlands," Global CCS Institute, 2010; https://www.globalccsinstitute.com/archive/hub/publications/8172/barendrecht-ccs-project-case-study.pdf.

128 **In 2019, 626 organizations** Green New Deal Letter to Congress, January 10, 2019; https://www.scribd.com/document/397201459/Green-New-Deal-Letter-to-Congress.

129 **"Christmas trees"** "An assembly of valves, fittings, chokes, and gauges used in monitoring and controlling producing, injection, and inactive wells. The Christmas tree is assembled at the top of the well starting with the uppermost flange of the tubing head." "Oil and Gas Drilling Glossary," International Association of Drilling Contractors; https://iadclexicon.org/christmas-tree/.

Chapter 10: Stoned

132 **Iceland's second-largest power plant** Wikipedia, accessed August 28, 2022; https://en.wikipedia.org/wiki/List_of_power_stations_in_Iceland.

133 **"The Geysers"** Calpine, "The Geysers"; https://geysers.com/geothermal.

133 **provides enough electricity to power San Francisco** "Good Question: How Much Electricity Does San Francisco Use Daily?" "18,000 MW hr of Elcctricity Each Day," CBS News; https://www.cbsnews.com/sanfrancisco/news/good-question-how-much-electricity-does-san-francisco-use-daily/. Geysers' annual generation of 6516 GWh in 2018, or 18 GWh per day; https://en.wikipedia.org/wiki/List_of_geothermal_power_stations.

134 **She and her colleagues designed a new approach** B. Sigfusson et al., "Solving the Carbon-Dioxide Buoyancy Challenge: The Design and Field Testing of a Dissolved CO_2 Injection System," *International Journal of Greenhouse Gas Control* 37 (2015): 213–19; https://doi.org/10.1016/j.ijggc.2015.02.022.

134 **Basalt is one of the most common rocks on Earth and it underlies much of Iceland** "An alternative is to inject CO2 into basaltic rocks, which contain up to 25% by weight of calcium, magnesium, and iron. Basaltic rocks are highly reactive and are one of the most common rock types on Earth, covering ~10% of continental surface area and most of the ocean floor." J. M. Matter et al., "Rapid Carbon Mineralization for Permanent Disposal of Anthropogenic Carbon Dioxide Emissions," *Science* 352 (2016): 1312–14; https://www.science.org/doi/10.1126/science.aad8132.

135 **More than 95 percent of the carbon dioxide injected below the Carbfix site** "We find that over 95% of the CO2 injected into the Carbfix site in Iceland was mineralized to carbonate minerals in less than 2 years. This result contrasts with the common view that the immobilization of CO2 as carbonate minerals within geologic reservoirs takes several hundreds to thousands of years. Our results, therefore, demonstrate that the safe long-term storage of anthropogenic CO2 emissions through mineralization can be far faster than previously postulated." J. M. Matter et al., "Rapid Carbon Mineralization for Permanent Disposal of Anthropogenic Carbon Dioxide Emissions," *Science*

352 (2016): 1312–14; https://www.science.org/doi/10.1126 /science.aad8132.

135 **the submersible pump in the injection well broke** "This 550-day limit also coincides with the breakdown of the submersible pump in HN04 monitoring well, which resulted in a 3-month gap in the subsequent monitoring data. The pump was clogged and coated with calcite." J. M. Matter et al., "Rapid Carbon Mineralization for Permanent Disposal of Anthropogenic Carbon Dioxide Emissions," *Science* 352 (2016): 1312–14; https://www.science .org/doi/10.1126/science.aad8132.

136 **Oklahoma suddenly had more earthquakes than California** Ivan Wong, "Induced Seismicity Primer Update," Oklahoma State Oil & Gas Regulatory Exchange, 2020; https://iogcc.ok.gov /sites/g/files/gmc836/f/wong_-_the_exchange_-_induced _seismicity_primer_update.pdf.

136 **A magnitude 5.7 earthquake near Prague, Oklahoma** D. F. Sumy et al., "Observations of Static Coulomb Stress Triggering of the November 2011 M5.7 Oklahoma Earthquake Sequence," *Journal of Geophysical Research: Solid Earth* 119 (2014): 1904–23; https://doi.org/10.1002/2013JB010612.

136 **Oklahoma implemented a "traffic light" system** Tim Baker, Oklahoma Corporation Commission's Oil and Gas Conservation Division, 2015; https://oklahoma.gov/content/dam/ok/en/occ /documents/ajls/news/archived-news/2015/03-25-15-me dia-advisory-tl-and-related-documents.pdf.

137 **the company is exploring using seawater for pumping carbon dioxide underground** "Carbfix tests using seawater to mineralize CO2 at Helguvík, Iceland," August 12, 2022; https://www.carbfix. com/carbfix-tests-using-seawater.

138 **the United States has only about five thousand miles of CO_2 pipelines** "Are CO2 Pipelines Regulated? By Whom?," Institute for Energy Research, March 8, 2023; https://www.institute-

forenergyresearch.org/regulation/are-co2-pipelines-regulated-by
-whom/.

138 **"Coda Terminal"** "The Biggest EU Grant to Iceland So Far Goes
to Carbfix," Northstack, July 19, 2022, https://www.northstack.is
/the-biggest-eu-grant-to-iceland-so-far-goes-to-carbfix/; "Carbfix's Coda Terminal Awarded Large EU Grant," Carbfix, July 12,
2022, https://www.carbfix.com/awarded-large-eu-grant.

138 **One recent project** Geothermal Emission Control or GECO; information on GECO can be found here: https://geco-h2020.eu.

138 **Carbfix's process costs less than $25 per metric ton** "Cost of
industrial scale Carbfix operations at Hellisheiði are less than $25/
ton, which is competitive with price of carbon allowances on the
ETS market and cheaper than alternative CCS methods." https://
www.carbfix.com/energy.

142 **Climeworks opened the first commercial air-capture
plant for CO$_2$ in Switzerland in 2017** "Commercial Operations of Climeworks' 1st Gen Technology Are Completed,"
Climeworks, October 20, 2022; https://climeworks.com/news
/climeworks-completes-commercial-operations-in-hinwil.

142 **Glen Peters coauthored a 2016 paper** K. Anderson and G. Peters, "The Trouble with Negative Emissions," *Science* 354 (2016):
182–83; https://doi.org/10.1126/science.aah4567.

Chapter 11: RePeat

145 **Rewilding** https://rewildingeurope.com/what-is-rewilding/.

147 **the gray heron is extending its range northward** M. Seaberg and
D Main, "Gray Heron Seen for First Time in Contiguous U.S., as Species Expands Range," *National Geographic*, 2020; https://www
.nationalgeographic.com/animals/article/gray-heron-sighting
-united-states-expanding-range.

148 **I'd read about such burst craters** Richard Gray, "The Mystery of Siberia's Exploding Craters," BBC Future, December 1, 2020; https://www.bbc.com/future/article/20201130-climate-change -the-mystery-of-siberias-explosive-craters.

154 **One team titled their recent paper** A. Günther et al., "Prompt Rewetting of Drained Peatlands Reduces Climate Warming Despite Methane Emissions," *Nature Communications* 11 (2020): 1644; https://www.nature.com/articles/s41467-020-15499-z.

154 **Other multiyear measurements in restored peatlands** K. Hemes et al., "A Biogeochemical Compromise: The High Methane Cost of Sequestering Carbon in Restored Wetlands," *Geophysical Research Letters* 45 (2018): 6081–91; https://doi .org/10.1029/2018GL077747.

Chapter 12: X-Methane

157 **four hundred billion tons' worth** G. Hugelius et al., "Large Stocks of Peatland Carbon and Nitrogen Are Vulnerable to Permafrost Thaw," *Proceedings of the National Academy of Sciences USA* 117 (2020): 20438–46; https://doi.org/10.1073/pnas.1916387117.

158 **studying new technologies and outlining benchmarks for progress** R. B. Jackson et al., "Atmospheric Methane Removal: A Research Agenda," *Philosophical Transactions of the Royal Society A* 379 (2021): 20200454, https://doi.org/10.1098 /rsta.2020.0454; R. B. Jackson et al., "Methane Removal and Atmospheric Restoration," *Nature Sustainability* 2 (2019): 436–38; https://www.nature.com/articles/s41893-019-0299-x.

158 **Metal catalysts** H. M. Rhoda et al., "Second-Sphere Lattice Effects in Copper and Iron Zeolite Catalysis," *Chemical Reviews* 122 (2022): 12207–43, https://doi.org/10.1021/acs.chem rev.1c00915; R. J. Brenneis et al., "Atmospheric- and Low-Level

Methane Abatement via an Earth-Abundant Catalyst, *ACS Environmental Au* 2 (2022): 223–31, https://pubs.acs.org/doi/pdf /10.1021/acsenvironau.1c00034.

159 **So have some light-activated "photocatalysts"** X. Chen et al. "Photocatalytic Oxidation of Methane over Silver Decorated Zinc Oxide Nanocatalysts," *Nature Communications* 7 (2016): 12273; https://www.nature.com/articles/ncomms12273/.

159 **Researchers in Mary Lidstrom's lab** L. He, J. D. Groom, E. H. Wilson, and M. E. Lidstrom, "A Methanotrophic Bacterium to Enable Methane Removal for Climate Mitigation," *Proceedings of the National Academy of Sciences, USA* 120 (2023): e2310046120; https://doi.org/10.1073/pnas.2310046120.

160 **This hydroxyl radical is sometimes called nature's detergent** D. Stone, L. K. Whalley, and D. E. Heard, "Tropospheric OH and HO_2 Radicals: Field Measurements and Model Comparisons," *Chemical Society Reviews* 41 (2012): 6348–404; https://doi .org/10.1039/C2CS35140D.

161 **Iron-rich dust blowing from the Sahara Desert** M. M. J. W. van Herpen et al., "Photocatalytic Chlorine Atom Production on Mineral Dust–Sea Spray Aerosols over the North Atlantic," *Proceedings of the National Academy of Sciences, USA* 120 (2023): e2303974120; https://doi.org/10.1073/pnas.2303974120.

162 **it needs much more research** R. B. Jackson et al., "Atmospheric Methane Removal: A Research Agenda," *Philosophical Transactions of the Royal Society A* 379 (2021): 20200454; https://doi .org/10.1098/rsta.2020.0454.

Chapter 13: Implausible Deniability

166 **If the thousands of fossil power plants on Earth today** D. Tong et al., "Committed Emissions from Existing Energy Infrastructure

Jeopardize 1.5°C Climate Target," *Nature* 572 (2019): 373–77; https://doi.org/10.1038/s41586-019-1364-3.

166 **In fact, more than half of known oil, gas, and coal reserves** D. Welsby et al., "Unextractable Fossil Fuels in a 1.5°C World," *Nature* 597 (2021): 230–34; https://doi.org/10.1038/s41586-021-03821-8.

166 **Ikard was responding to a newly released report** Environmental Pollution Panel, *Restoring the Quality of Our Environment: Report* (Washington, D.C.: President's Science Advisory Committee, The White House, 1965), 111–33; https://www.documentcloud.org/documents/3227654-PSAC-1965-Restoring-the-Quality-of-Our-Environment.

167 **that documented the buildup of fossil carbon dioxide in the atmosphere** B. Franta, "Early Oil Industry Knowledge of CO2 and Global Warming," *Nature Climate Change* 8 (2018): 1024–25, https://www.nature.com/articles/s41558-018-0349-9; F. N. Ikard, "Meeting the Challenges of 1966," *Annual Meeting of the American Petroleum Institute 1965* (API, 1965), 12–15.

167 **Harvard historian Naomi Oreskes** G. Supran, S. Rahmstorf, and N. Oreskes, "Assessing ExxonMobil's Global Warming Projections," *Science* 379 (2023): 6628, https://doi.org/10.1126/science.abk0063; see also https://news.harvard.edu/gazette/story/2023/01/harvard-led-analysis-finds-exxonmobil-internal-research-accurately-predicted-climate-change/.

167 **Frank Sprow, drafted a memo for colleagues** C. M. Matthews and C. Eaton, "Inside Exxon's Strategy to Downplay Climate Change," *Wall Street Journal*, September 14, 2023; https://www.wsj.com/business/energy-oil/exxon-climate-change-documents-e2e9e6af.

168 **spending $60 billion to purchase Pioneer Natural Resources** "ExxonMobil Announces Merger with Pioneer Natural Resources in an All-Stock Transaction," ExxonMobil, October 11, 2023; https://investor.exxonmobil.com/news-events/press-releases/detail/1147/exxonmobil-announces-merger-with-pioneer-natural-resources.

168 **The Permian Basin yields one-fifth of all gas produced in the United States** June 13, 2023, U.S. Energy Information Administration, Natural gas production in the Permian region established a new record in 2022; https://www.eia.gov/todayinenergy/detail.php?id=56800#.

168 **Keith McCoy, senior director for federal relations at Exxon-Mobil** https://www.youtube.com/watch?v=dNFBjcrU5Pc.

168 **Former senior vice president and president of Manville's Fiberglass Group Bill Sells** Bill Sells, "What Asbestos Taught Me About Managing Risk," *Harvard Business Review* (March-April 1994); https://hbr.org/1994/03/what-asbestos-taught-me-about-managing-risk.

169 **A recent article** W. F. Lamb et al., "Discourses of Climate Delay," *Global Sustainability* 3 (2020): e17; https://doi.org/10.1017/sus.2020.13.

170 **acquisition of oil giant Amoco** "BP and Amoco's Mega-merger Two Decades On," BP; https://www.bp.com/en/global/corporate/news-and-insights/reimagining-energy/20-year-anniversary-bp-amoco-merger.html.

170 **speech at Stanford in 2002** "Environmental and Social Review 2002," BP; https://www.bp.com/content/dam/bp/business-sites/en/global/corporate/pdfs/sustainability/archive/archived-reports-and-translations/2003-1998/environmental-and-social-report-2002.pdf.

170 **In one archived television ad** https://www.youtube.com/watch?v=ywrZPypqSB4.

171 **Advertising executive John Kenney** John Kenney, "Beyond Propaganda," op-ed, *New York Times*, August 14, 2006; https://www.nytimes.com/2006/08/14/opinion/14kenney.html.

171 **BP spent more than $10 billion** "BP Completes Purchase of BHP Assets in US Onshore," BP, October 31, 2018; https://www.bp.com/en/global/corporate/news-and-insights/press-releases/bp-completes-purchase-of-bhp-assets-in-us-onshore.html.

172 The BP press release "bpx energy," October 31, 2018; https://www.bp.com/en_us/united-states/home/news/press-releases/bp-completes-purchase-of-bhp-assets-in-us-onshore.html.

172 According to Martin Neubert "Ørsted's Renewable-Energy Transformation," McKinsey Sustainability, July 10, 2020; https://www.mckinsey.com/capabilities/sustainability/our-insights/orsteds-renewable-energy-transformation; see also https://orsted.com.

Chapter 14: Some Heroes to Zero

175 *Peromyia karstroemi* https://en.wikipedia.org/wiki/Peromyia.

175 *Campylomyza stegetfore* https://en.wikipedia.org/wiki/Campylomyza.

176 Jielkká-Rijmagåbbå forest https://eunis.eea.europa.eu/sites/SE0820722.

178 One 2019 story from the University of Würzburg reads "Scientists Alarmed by Bark Beetle Boom"; https://www.uni-wuerzburg.de/en/news-and-events/news/detail/news/scientists-alarmed-by-bark-beetle-boom/.

178 Another, from 2020 P. Jonsson, "The Bark-Beetle Invades Europe," *NordicWoodJournal*, March 3, 2020; https://nordicwoodjournal.com/editorial/bark-beetle-invades-europe/.

180 "New Green Hero" https://www.rollingstone.com/politics/politics-lists/the-fossil-fuel-resistance-meet-the-new-green-heroes-142112/rev-lennox-yearwood-jr-the-minister-174202/.

180 "Champion of Change" https://obamawhitehouse.archives.gov/champions/climate-faith-leaders/reverend-lennox-yearwood-jr.

180 Scientific papers K. E. Trenberth, C. A. Davis, and J. Fasullo, "Water and Energy Budgets of Hurricanes: Case Studies of Ivan and Katrina," *Journal of Geophysical Research Atmospheres* 112 (2007), https://doi.org/10.1029/2006JD008303; J. L. Irish et al.,

"Simulations of Hurricane Katrina (2005) Under Sea Level and Climate Conditions for 1900," *Climatic Change* 122 (2013): 635–49, https://link.springer.com/article/10.1007/s10584-013-1011-1.

182 **She and I have worked together to measure soil organic carbon** K. Georgiou et al., "Global Stocks and Capacity of Mineral-Associated Soil Organic Carbon," *Nature Communications* 13 (2022): 3797; https://www.nature.com/articles/s41467-022-31540-9.

183 **Alaska shattered its record for warmest year** A. Rubel and R. Thoman, "2016 Shatters Record for Alaska's Warmest Year," Climate.gov, January 10, 2017; https://www.climate.gov/news-features/features/2016-shatters-record-alaskas-warmest-year.

184 **Here was the group's statement in an open letter** https://signon.scientistrebellion.com.

184 **As described in a Department of Energy announcement** "President Biden invokes Defense Production Act to accelerate domestic manufacturing of clean energy," June 6, 2022; https://www.energy.gov/articles/president-biden-invokes-defense-production-act-accelerate-domestic-manufacturing-clean.

186 **Peter Kalmus** D. Boraks, "These Climate Scientists Feel 'A Higher Calling'—Civil Disobedience," WFAE, November 11, 2022; https://www.wfae.org/energy-environment/2022-11-11/these-climate-scientists-feel-a-higher-calling-civil-disobedience.

Chapter 15: Into the Clear

189 **Scientists using satellite data** L. Feng et al., "Tropical Methane Emissions Explain Large Fraction of Recent Changes in Global Atmospheric Methane Growth Rate," *Nature Communications* 13 (2022): 1378; https://www.nature.com/articles/s41467-022-28989-z.

189 **illegal gold mining is surging** S. Daley, "Peru Scrambles to Drive Out Illegal Gold Mining and Save Precious Land," *New York Times*, July 25, 2016, https://www.nytimes.com/2016/07/26 /world/americas/peru-illegal-gold-mining-latin-america.html; see also a collaboration between Conservación Amazónica, NASA, and Peru's Ministry of Environment: A. C. Keck, "Teaming Up to Track Illegal Amazon Gold Mines in Peru," NASA, January 26, 2021, https://www.nasa.gov/feature/teaming-up-to-track-illegal -amazon-gold-mines-in-peru.

189 **the mercury used to harvest the gold** J. R. Gerson, "Amazon Forests Capture High Levels of Atmospheric Mercury Pollution from Artisanal Gold Mining," *Nature Communications* 13 (2022): 559, https://www.nature.com/articles/s41467-022-27997-3; see also Z. Snowdon Smith, *Forbes*, January 28, 2022, https://www .forbes.com/sites/zacharysmith/2022/01/28/illegal-gold-min ing-causes-devastating-mercury-pollution-in-amazon-rainforest -study-says/?sh=1f75a9d2521d.

189 **Such "tipping points" may kill rainforests** C. E. Doughty et al., "Tropical Forests Are Approaching Critical Temperature Thresholds," *Nature* 621 (2023): 105–11; https://www.nature.com/articles /s41586-023-06391-z.

190 **showing a one-quarter increase in the maximum flooded area across the Amazon floodplain** A. S. Fleischmann et al., "Increased Floodplain Inundation in the Amazon since 1980," *Environmental Research Letters* 18 (2023): 034024; https://iop science.iop.org/article/10.1088/1748-9326/acb9a7/meta.

190 **Hoatzin are also the world's only flying cow** A.-D. G. Wright, K. S. Northwood, and N. E. Obispo, "Rumen-like Methanogens Identified from the Crop of the Folivorous South American Bird, the Hoatzin (*Opisthocomus hoazin*)," *ISME Journal* 3 (2009): 1120–26; https://www.nature.com/articles/ismej200941.

191 **extreme drought triggered by the 2015–2016 El Niño** J. C. Jiménez-Muñoz et al., "Record-Breaking Warming and Extreme Drought in the Amazon Rainforest During the Course of El Niño 2015–2016," *Scientific Reports* 6 (2016): 33130; https://www.na ture.com/articles/srep33130.

191 **temperatures suggested for cooking Atlantic salmon** "The Best Sous Vide Salmon Recipe," Sous Vide Ways, March, 28, 2020; https://sousvideways.com/best-sous-vide-salmon-recipe/.

191 **the highest surface ocean temperatures measured** E. Zerkel, "Ocean Heat Around Florida Is 'Unprecedented,' and Scientists Are Warning of Major Impacts," CNN, July 16, 2023; https://www.cnn.com/2023/07/12/us/florida-ocean-heat-coral-bleach ing-climate/index.html.

191 **the risks of overheating and exhaustion** L. Norman, "Florida's Ocean Water Temperatures Are High. But Are They too Hot for You to Swim?," *Palm Beach Post*, July 28, 2023; https://www .palmbeachpost.com/story/news/environment/2023/07/28 /florida-oceans-heat-wave-what-to-know-before-you-go-ex treme-heat-miami-florida-keys-coral-reefs/70483921007/.

191 **Diana Nyad, "What It's Like to Swim in an Ocean That's 100 De grees," *New York Times*, August 14, 2023; https://www.nytimes .com/2023/08/14/opinion/ocean-temperature-climate-change .html.

191 **Water levels in the Amazon system were lower than at any time since record-keeping began in the early 1900s** F. Mai sonnave, "In Brazil's Amazon, Rivers Fall to Record Low Lev els During Drought," Associated Press, October 16, 2023; https://apnews.com/article/amazon-brazil-negro-river-drought -manaus-88421d0cbadca79007b4485cbf497c11.

191 **Brazil's minister for the environment, Marina Silva, said** "Brazil Sets Up Task Force for Unprecedented Drought in Am-

azon—Minister," Reuters, September 27, 2023; https://www
.usnews.com/news/world/articles/2023-09-27/brazil-sets-up
-task-force-for-unprecedented-drought-in-amazon-minister.

192 **he measured water temperatures at an astounding 105°F**
Fleischmann wrote me, "I measured 39.1°C in a whole 2 m deep
water column in Lake Tefé, and 40.9°C in a whole 1 m deep portion
of the lake. But I know it reached even higher in other upstream
reaches (even shallower, like 30 cm) but we couldn't get there
to measure. I made a quick search and talked to many scientists
already, I think nobody has ever measured anything like that (2 m
deep of >39 °C) in tropical lakes."

192 **more than seven thousand fires raged across Amazonas state**
E. Barros, "'Without water, there is no life': Drought in Brazil's Amazon Is Sharpening Fears for the Future," Associated Press, October 8,
2023; https://apnews.com/article/brazil-amazon-drought-environ
ment-climate-rivers-4152b5e288e1b85abc4e5d4af64db67d.

192 **News photos displayed dead dolphins** "Severe Drought
Devastates Amazon, Rare Dolphins Die as Water Temperatures Soar," Reuters, October 5, 2023; https://www.reuters
.com/pictures/brazils-amazon-drought-affects-locals-access
-food-water-2023-10-03/MXP75LE4BBO7TGIJ7OS6KFWX
UU/?taid=651f07d1ce2bfe00016da6ae.

192 **"Without water, there is no life," said a local fisherman**
E. Barros, "'Without water, there is no life': Drought in Brazil's Amazon Is Sharpening Fears for the Future," Associated
Press, October 8, 2023; https://apnews.com/article/brazil-ama
zon-drought-environment-climate-rivers-4152b5e288e1b85ab
c4e5d4af64db67d.

193 **Europe's hottest-ever summer that killed sixty thousand people** J. Ballester et al., "Heat-Related Mortality in Europe During
the Summer of 2022," *Nature Medicine* 29 (2023): 1857–66;
https://www.nature.com/articles/s41591-023-02419-z.

193 **the sweltering heat wave in the northwestern United States and Canada** "Monday's record-setting temperatures broke Sunday's record-setting high of 112 degrees. Sunday's high had broken the 108 degree–record set Saturday, which broke the previous high of 107, first set in 1965." Jamie Goldberg and Jayati Ramakrishnan, "Portland Records All-Time High Temperature of 116, Setting New Record for Third Day in a Row," OregonLive.com, June 29, 2021; https://www.oregonlive.com/weather/2021/06/portland-records-all-time-high-temperature-of-113-setting-new-record-for-third-day-in-a-row.html.

193 **Lytton, British Columbia, set Canada's highest temperature ever recorded** S. Moon and R. Riess, "More Than 230 Deaths Reported in British Columbia Amid Historic Heat Wave," CNN, June 30, 2021; "Lytton, British Columbia, hit 121 degrees on Tuesday—the highest temperature ever recorded in Canada," https://www.cnn.com/2021/06/29/americas/canada-heat-wave-deaths/index.html; see also CTV News, "'It'll Be a Total Rebuild': Lytton, B.C., Mayor Describes Work Underway and Next Steps," https://bc.ctvnews.ca/it-ll-be-a-total-rebuild-lytton-b-c-mayor-describes-work-underway-and-next-steps-1.5494709; "'From what I can see in town there's only a few houses left so it'll be a total rebuild,' Jan Polderman said."

193 **the single hottest month ever worldwide—July** NASA, August 14, 2023, "NASA Clocks July 2023 as Hottest Month on Record Ever Since 1880," https://climate.nasa.gov/news/3279/nasa-clocks-july-2023-as-hottest-month-on-record-ever-since-1880/.

193 **hottest year on record—2023** R. Ramirez, "2023 Will Officially Be the Hottest Year on Record, Scientists Report," CNN, December 6, 2023, https://www.cnn.com/2023/12/06/climate/2023-hottest-year-climate/index.html; see also European Commission, https://climate.copernicus.eu/record-warm-november-consolidates-2023-warmest-year.

193 **Forty-six million Canadian acres burned by October** Fire Statistics, Canadian Interagency Forest Fire Centre; https://ciffc.net/statistics.

193 **Dangerous wildfire pollution choked people from Ottawa to Miami** J. Nix, *Forbes*, July 25, 2023; https://www.forbes.com/sites/jessicanix/2023/07/25/three-major-us-cities-ranked-in-worlds-top-20-for-worst-air-quality/?sh=765e03497622.

193 **A longtime wildland firefighter and Canadian fire chief** D. Wallace-Wells, *New York Times*, October 24, 2023; https://www.nytimes.com/2023/10/24/magazine/canada-wildfires.html.

194 **I was quoted in the *Washington Post* saying** C. Mooney and B. Dennis, "Global Greenhouse Gas Emissions Will Hit Another Record High This Year, Experts Project," *Washington Post*, December 3, 2019; https://www.washingtonpost.com/climate-environment/2019/12/03/global-greenhouse-gas-emissions-will-hit-yet-another-record-high-this-year-experts-project/.

194 **I've called the *Hellocene*** Rob Jackson, *Scientific American*, November 26, 2018; https://blogs.scientificamerican.com/observations/youve-heard-of-the-anthropocene-welcome-to-the-hellocene/.

194 **the "emergence of heat and humidity too severe for human tolerance"** C. Raymond, T. Matthews, and R. M. Horton "The Emergence of Heat and Humidity Too Severe for Human Tolerance," *Science Advances* 6 (2020); https://doi.org/10.1126/sciadv.aaw1838.

194 **More than thirty-five years ago, he concluded** Transcript of James Hansen's U.S. Senate Testimony, June 23, 1988; https://www.pulitzercenter.org/sites/default/files/june_23_1988_senate_hearing_1.pdf.

194 **"We are damned fools"** O. Milman, "'We are damned fools': Scientist Who Sounded Climate Alarm in 80s Warns of Worse to

Come," *Guardian*, July 19, 2023; https://www.theguardian.com
/environment/2023/jul/19/climate-crisis-james-hansen-scientist
-warning.

194 **The landmark decision struck down two Montana laws** S. Bookman, "Held v. Montana: A Win for Young Climate Advocates and What It Means for Future Litigation," August 30, 2023; https://eelp.law.harvard.edu/2023/08/held-v-montana/.

194 **The judge presiding over the case determined** J. Powel, "Group Behind Montana Youth Climate Lawsuit Has Lawsuits in 3 Other State Courts: What to Know," *USA Today*, August 15, 2023; https://www.usatoday.com/story/news/nation/2023/08/15/youth-climate-lawsuits-hawaii-virginia-utah-mirror-montana/70592574007/.

195 **One of the plaintiffs, Grace Gibson-Snyder, listened early in her life** "Meet One of the Teens Suing Montana over Climate Crisis. She Says Planet's Future Is at Stake," democracynow.org, June 29, 2023; https://www.democracynow.org/2023/6/29/montana_climate_trial.

INDEX

INDEX